Shock-Wave Solutions of the Einstein Equations with Perfect Fluid Sources: Existence and Consistency by a Locally Inertial Glimm Scheme

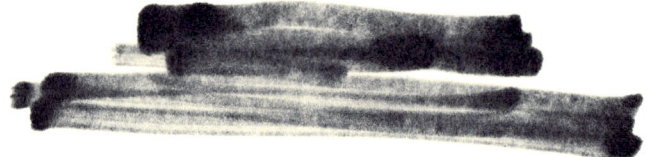

Number 813

Shock-Wave Solutions of the Einstein Equations with Perfect Fluid Sources: Existence and Consistency by a Locally Inertial Glimm Scheme

Jeff Groah
Blake Temple

American Mathematical Society
Providence, Rhode Island

2000 *Mathematics Subject Classification.* Primary 35L65, 35L67, 83C05.

Library of Congress Cataloging-in-Publication Data

Groah, Jeff.

Shock-wave solutions of the Einstein equations with perfect fluid sources : existence and consistency by a locally inertial Glimm scheme / Jeff Groah, Blake Temple.
    p. cm. — (Memoirs of the American Mathematical Society, ISSN 0065-9266 ; no. 813)
"Volume 172, number 813 (second of 4 numbers)."
Includes bibliographical references.
ISBN 0-8218-3553-X (alk. paper)
    1. Conservation laws (Physics) 2. Einstein field equations. 3. Shock waves. 4. Singularities (Mathematics) 5. General relativity (Physics) I. Temple, Blake. II. Title. III. Series.

QA3.A57 no. 813
[QC793.3.C58]
510 s—dc22
[531′.1133]
                            2004054527

Memoirs of the American Mathematical Society

This journal is devoted entirely to research in pure and applied mathematics.

Subscription information. The 2004 subscription begins with volume 167 and consists of six mailings, each containing one or more numbers. Subscription prices for 2004 are $583 list, $466 institutional member. A late charge of 10% of the subscription price will be imposed on orders received from nonmembers after January 1 of the subscription year. Subscribers outside the United States and India must pay a postage surcharge of $31; subscribers in India must pay a postage surcharge of $43. Expedited delivery to destinations in North America $35; elsewhere $130. Each number may be ordered separately; *please specify number* when ordering an individual number. For prices and titles of recently released numbers, see the New Publications sections of the *Notices of the American Mathematical Society*.

Back number information. For back issues see the *AMS Catalog of Publications*.

Subscriptions and orders should be addressed to the American Mathematical Society, P. O. Box 845904, Boston, MA 02284-5904, USA. *All orders must be accompanied by payment.* Other correspondence should be addressed to 201 Charles Street, Providence, RI 02904-2294, USA.

Copying and reprinting. Individual readers of this publication, and nonprofit libraries acting for them, are permitted to make fair use of the material, such as to copy a chapter for use in teaching or research. Permission is granted to quote brief passages from this publication in reviews, provided the customary acknowledgment of the source is given.

Republication, systematic copying, or multiple reproduction of any material in this publication is permitted only under license from the American Mathematical Society. Requests for such permission should be addressed to the Acquisitions Department, American Mathematical Society, 201 Charles Street, Providence, Rhode Island 02904-2294, USA. Requests can also be made by e-mail to `reprint-permission@ams.org`.

Memoirs of the American Mathematical Society is published bimonthly (each volume consisting usually of more than one number) by the American Mathematical Society at 201 Charles Street, Providence, RI 02904-2294, USA. Periodicals postage paid at Providence, RI. Postmaster: Send address changes to Memoirs, American Mathematical Society, 201 Charles Street, Providence, RI 02904-2294, USA.

© 2004 by the American Mathematical Society. All rights reserved.
This publication is indexed in *Science Citation Index*®, *SciSearch*®, *Research Alert*®, *CompuMath Citation Index*®, *Current Contents*®/*Physical, Chemical & Earth Sciences*.
Printed in the United States of America.

∞ The paper used in this book is acid-free and falls within the guidelines established to ensure permanence and durability.
Visit the AMS home page at `http://www.ams.org/`

10 9 8 7 6 5 4 3 2 1   09 08 07 06 05 04

Contents

1 **Introduction**   1
   1.1 The Proof Strategy 5
   1.2 A Locally Inertial Glimm Scheme 9
   1.3 The Smoothness Class of the Metric 9

2 **Preliminaries**   11

3 **The Fractional Step Scheme.**   17

4 **The Riemann Problem Step**   25

5 **The ODE Step**   34

6 **Estimates for the ODE step**   40

7 **Analysis of the Approximate Solutions**   46

8 **The Elimination of Assumptions**   60

9 **Convergence**   73

Abstract

We demonstrate the consistency of the Einstein equations at the level of shock-waves by proving the existence of shock wave solutions of the spherically symmetric Einstein equations for a perfect fluid, starting from initial density and velocity profiles that are only locally of bounded total variation. For these solutions, the components of the gravitational metric tensor are only Lipschitz continuous at shock waves, and so it follows that these solutions satisfy the Einstein equations, as well as the relativistic compressible Euler equations, only in the weak sense of the theory of distributions. The analysis introduces a *locally inertial* Glimm scheme that exploits the locally flat character of spacetime, and relies on special properties of the relativistic compressible Euler equations when $p = \sigma^2 \rho$, $\sigma \equiv const$.

AMS Subject Classification Numbers: 35L65,35L67,83C05
Key Words and Phrases: Shock Waves, Glimm Scheme, General Relativity

SHOCK-WAVE SOLUTIONS OF THE EINSTEIN EQUATIONS WITH PERFECT FLUID SOURCES:

EXISTENCE and CONSISTENCY

by a

LOCALLY INERTIAL GLIMM SCHEME

Jeff Groah[1] *Blake Temple*[2]

1 Introduction

In General Relativity, a time dependent, spherically symmetric gravitational metric can, (under generic conditions), be transformed over to standard Schwarzschild coordinates $\mathbf{x} = (x^0, x^1, x^2, x^3) \equiv (t, r, \theta, \phi)$, where the metric takes the canonical form, [30],

$$ds^2 = -Adt^2 + Bdr^2 + r^2\left(d\theta^2 + \sin^2\theta d\phi^2\right), \tag{1.1}$$

where the metric components A and B are assumed to be functions of (t, r), $A = A(r, t)$, $B = B(r, t)$. In this paper we establish the consistency of the Einstein equations at the level of shock waves by proving the existence of shock-wave solutions of the Einstein-Euler equations for gravitational metrics of form (1.1), for general initial density and velocity profiles that are only locally functions of bounded total variation. The solutions are defined outside a ball of fixed total mass[4], existence is proved up to some positive time $T > 0$[5], and the total mass at $r \to \infty$ is shown to be constant throughout the time interval $[0, T)$. To keep the analysis as simple as possible, we assume the equation of state $p = \sigma^2\rho$, $p = pressure$, $\rho = density$, where σ, the sound

---

[1] Department of Mathematics, California State University at Monterey Bay

[2] Department of Mathematics, University of California at Davis

[3] Received by the editor March 14, 2002. The second author supported in part by NSF Applied Mathematics Grant Number DMS-010-2493 and by the Institute of Theoretical Dynamics, UC-Davis.

[4] This removes the possibility of waves perfectly focused on the origin, that can amplify to infinity.

[5] One can only expect a finite time existence result because, in standard Schwarzschild coordinates, solutions blow up at black hole singularities, i.e., $B = \frac{1}{1-\frac{2M}{r}} \to \infty$ at a black hole, and black holes can form in finite time

speed, is assumed to be constant[6]. For these solutions, the fluid variables ρ, p and velocity $\mathbf{w}$, are in general *discontinuous*, and the metric components A and B are only *Lipschitz continuous functions*, at the shock waves. Since the Einstein equations involve second derivatives of A and B, (c.f. (1.5) below), it follows that these solutions satisfy the Einstein equations only in the weak sense of the theory of distributions. Thus our theorem establishes the consistency of the initial value problem for the Einstein equations at the weaker level of shock-waves.

We now discuss the main theorem in detail. In standard Schwarzschild coordinates, the Einstein equations of General Relativity reduce to the following system of four partial differential equations, (see (3.20)-(3.23) of [10]),

$$\frac{A}{r^2 B}\left\{r\frac{B'}{B} + B - 1\right\} = \kappa A^2 T^{00} \tag{1.2}$$

$$-\frac{B_t}{rB} = \kappa AB T^{01} \tag{1.3}$$

$$\frac{1}{r^2}\left\{r\frac{A'}{A} - (B-1)\right\} = \kappa B^2 T^{11} \tag{1.4}$$

$$-\frac{1}{rAB^2}\{B_{tt} - A'' + \Phi\} = \frac{2\kappa r}{B} T^{22}, \tag{1.5}$$

where the quantity Φ in the last equation is given by,

$$\Phi = -\frac{BA_t B_t}{2AB} - \frac{B}{2}\left(\frac{B_t}{B}\right)^2 - \frac{A'}{r} + \frac{AB'}{rB} + \frac{A}{2}\left(\frac{A'}{A}\right)^2 + \frac{A}{2}\frac{A'}{A}\frac{B'}{B}.$$

Here "prime" denotes $\partial/\partial r$, "dot" denotes $\partial/\partial t$, $\kappa = \frac{8\pi\mathcal{G}}{c^4}$ is the coupling consant, $\mathcal{G}$ is Newton's gravitational constant, c is the speed of light, T^{ij}, $i, j = 0, ..., 3$ are the components of the stress energy tensor, and $A \equiv A(r, t)$, $B \equiv B(r, t)$ denote the components of the gravitational metric tensor (1.1). The mass function $M(r, t)$ is defined through the identity

---

[6]This simplifying assumption, as well as insuring that wave speeds are bounded by the speed of light for arbitrarily strong shock waves, also prevents the formation of vacuum states. Moreover, the analysis exploits the existence of a *Nishida functional*, that is nonincreasing on weak solutions of the compressible Euler equations in flat spacetime, and only exists when $p = \sigma^2 \rho$, [20]. The existence of the Nishida functional in the relativistic regime was discovered by Smoller and Temple in [24].

$$B = \left(1 - \frac{2M}{r}\right)^{-1}, \tag{1.6}$$

and $M = M \equiv M(r,t)$ is interpreted as the mass inside radius r at time t. In terms of the variable M, equations (1.2) and (1.3) are equivalent to

$$M' = \tfrac{1}{2}\kappa r^2 A T^{00}, \tag{1.7}$$

and

$$\dot{M} = -\tfrac{1}{2}\kappa r^2 A T^{01}, \tag{1.8}$$

respectively. In the case when the stress tensor T is taken to be the stress tensor for a perfect fluid,

$$T^{ij} = (\rho c^2 + p)w^i w^j + p g^{ij}, \quad i,j = 0, ..., 3, \tag{1.9}$$

system (1.2)-(1.5) gives the spherically symmetric version of the Einstein-Euler equations,

$$G^{ij} = \kappa T^{ij}, \quad T^{ij} = (\rho c^2 + p)w^i w^j + p g^{ij}, \quad i,j = 0, ..., 3, \tag{1.10}$$

where G is the Einstein curvature tensor, ρc^2 is the energy density, p is the pressure, and $\mathbf{w}$ is the four velocity of the fluid. That is, system (1.2)-(1.5) is obtained from (1.10) by substituting for G_{ij} the components of the Einstein curvature tensor associated with the metric ansatz (1.1). System (1.10) describes the coupling of a compressible fluid to the gravitational metric tensor according Einstein's theory of general relativity.

The components T^{ij} satisfy

$$T^{00} = \frac{1}{A} T^{00}_M, \tag{1.11}$$

$$T^{01} = \frac{1}{\sqrt{AB}} T^{01}_M, \tag{1.12}$$

$$T^{11} = \frac{1}{B} T^{11}_M, \tag{1.13}$$

where T^{ij}_M denote the components of T in flat Minkowski spacetime. Assuming the equation of state

$$p = \sigma^2 \rho, \quad 0 < \sigma < c, \tag{1.14}$$

$\sigma \equiv constant$, and assuming that **w** is radial, the components of T_M can be written in the form

$$T_M^{00} = \frac{c^4 + \sigma^2 v^2}{c^2 - v^2}\rho, \tag{1.15}$$

$$T_M^{01} = \frac{c^2 + \sigma^2}{c^2 - v^2}cv\rho, \tag{1.16}$$

$$T_M^{11} = \frac{v^2 + \sigma^2}{c^2 - v^2}\rho c^2, \tag{1.17}$$

c.f., [24, 10]. Here v, taken in place of **w**, denotes the fluid velocity as measured by an observer fixed with respect to the radial coordinate r. It follows from (1.7) together with (1.15)-(1.17) that, if $r \geq r_0 > 0$, then

$$M(r,t) = M(r_0, t) + \frac{\kappa}{2}\int_{r_0}^{r} T_M^{00}(r,t)r^2\, dr; \tag{1.18}$$

it follows from (1.10) together with (1.15)-(1.17) that the scalar curvature R is proportional to the density,

$$R = (c^2 - 3\sigma^2)\rho; \tag{1.19}$$

and it follows directly form (1.15)-(1.17) that

$$|T_M^{01}| < T_M^{00}, \tag{1.20}$$

$$\frac{\sigma^2}{c^2+\sigma^2}T_M^{00} < T_M^{11} < T_M^{00}. \tag{1.21}$$

Equations (1.1)-(1.21) define the simplest possible setting for shock wave propagation in the Einstein equations.

For our theorem, assume the initial boundary conditions

$$\rho(r,0) = \rho_0(r), \quad v(r,0) = v_0(r), \quad for\ r > r_0,$$
$$M(r_0, t) = M_{r_0}, \quad v(r_0, t) = 0, \quad for\ t \geq 0, \tag{1.22}$$

where r_0 and M_{r_0} are positive constants, and assume the no black hole and finite total mass conditions,

$$\frac{2M(r,t)}{r} < 1, \quad \lim_{r \to \infty} M(r,t) = M_\infty < \infty, \tag{1.23}$$

hold at $t = 0$. For convenience, assume further that

$$\lim_{r \to \infty} r^2 T_M^{00}(r,t) = 0, \tag{1.24}$$

holds at $t = 0$, c.f., (1.18), (1.23). The main result of this paper can be stated as follows:

Theorem 1 *Assume that the initial boundary data satisfy (1.22)-(1.24), and assume that there exist positive constants L, V and $\bar{v}$ such that the initial velocity and density profiles $v_0(r)$ and $\rho_0(r)$ satisfy*

$$TV_{[r,r+L]} \ln \rho_0(\cdot) < V, \quad TV_{[r,r+L]} \ln \left(\frac{c + v_0(\cdot)}{c - v_0(\cdot)} \right) < V, \quad |v_0(r)| < \bar{v} < c, \tag{1.25}$$

for all $r_0 \leq r < \infty$, where $TV_{[a,b]} f(\cdot)$ denotes the total variation of the function f over the interval $[a,b]$. Then a bounded weak (shock wave) solution of (1.2)-(1.5), satisfying (1.22) and (1.23), exists up to some positive time $T > 0$. Moreover, the metric functions A and B are Lipschitz continuous functions of (r,t), and (1.25) continues to hold for $t < T$ with adjusted values for V and $\bar{v}$ that are determined from the analysis.

Note that the theorem allows for arbitrary numbers of interacting shock waves, of arbitrary strength. Note that by (1.2), (1.4), the metric components A and B will be no smoother than Lipschitz continuous when shocks are present, and thus since (1.5) is second order in the metric, it follows that (1.5) is only satisfied in the weak sense of the theory of distributions. Note finally that $\lim_{r \to \infty} M(r,t) = M_\infty$ is a *non-local* condition.

1.1 The Proof Strategy

In previous work [10], the authors show that when the metric components A and B are Lipschitz continuous, and T is bounded in L^∞, (when viewed as functions of the coordinate variables (t, r, θ, ϕ)), system (1.2)-(1.5) is weakly equivalent to the following system of equations obtained by replacing (1.3) and (1.5) with the 0- and 1-components of (covariant) $DivT = 0$,

$$\{T_M^{00}\}_{,0} + \left\{ \sqrt{\frac{A}{B}} T_M^{01} \right\}_{,1} = -\frac{2}{x} \sqrt{\frac{A}{B}} T_M^{01}, \tag{1.26}$$

$$\{T_M^{01}\}_{,0} + \left\{ \sqrt{\frac{A}{B}} T_M^{11} \right\}_{,1} = -\frac{1}{2} \sqrt{\frac{A}{B}} \left\{ \frac{4}{x} T_M^{11} + \frac{(B-1)}{x} (T_M^{00} - T_M^{11}) \right\} \tag{1.27}$$

$$+2\kappa x B(T_M^{00}T_M^{11} - (T_M^{01})^2) - 4xT^{22}\},$$

$$\frac{B'}{B} = -\frac{(B-1)}{x} + \kappa x B T_M^{00}, \tag{1.28}$$

$$\frac{A'}{A} = \frac{(B-1)}{x} + \kappa x B T_M^{11}. \tag{1.29}$$

This is the system of equations that we work with here. (Cf. (4.67), (4.68) together with (3.20), (3.22) of [10].) Here, ",i" denotes $\partial/\partial x^i$, and T_M is defined in (1.15)-(1.17).

System (1.26),(1.27),(1.28),(1.29) forms a system of conservation laws with source terms which we write in the compact form, (c.f. [10]),

$$u_t + f(\mathbf{A}, u)_x = g(\mathbf{A}, u, x), \tag{1.30}$$
$$\mathbf{A}' = h(\mathbf{A}, u, x), \tag{1.31}$$

where

$$u = (T_M^{00}, T_M^{01}) \equiv (u^0, u^1), \tag{1.32}$$
$$\mathbf{A} = (A, B), \tag{1.33}$$
$$f(\mathbf{A}, u) = \sqrt{\frac{A}{B}}\left(T_M^{01}, T_M^{11}\right), \tag{1.34}$$

and

$$g(\mathbf{A}, u, x) = \left(g^0(\mathbf{A}, u, x), g^1(\mathbf{A}, u, x)\right), \tag{1.35}$$

$$h(\mathbf{A}, u, x) = \left(h^0(\mathbf{A}, u, x), h^1(\mathbf{A}, u, x)\right), \tag{1.36}$$

where

$$g^0(\mathbf{A}, u, x) = -\frac{2}{x}\sqrt{\frac{A}{B}}T_M^{01}, \tag{1.37}$$

$$g^1(\mathbf{A}, u, x) = -\frac{1}{2}\sqrt{\frac{A}{B}}\left\{\frac{4}{x}T_M^{11} + \frac{(B-1)}{x}(T_M^{00} - T_M^{11})\right. \tag{1.38}$$
$$\left.+2\kappa x B(T_M^{00}T_M^{11} - (T_M^{01})^2) - 4xT^{22}\right\},$$

and

$$h^0(\mathbf{A}, u, x) = \frac{(B-1)A}{x} + \kappa x ABT_M^{11}, \qquad (1.39)$$

$$h^1(\mathbf{A}, u, x) = -\frac{(B-1)B}{x} + \kappa x B^2 T_M^{00}. \qquad (1.40)$$

The vector $h(\mathbf{A}, u, x)$ is just obtained by solving (1.2) and (1.4) for A' and B'. Note that we have set $x \equiv x^1 \equiv r$, and we use x in place of r in the analysis to follow since this is standard notation in the literature on hyperbolic conservation laws. Note also that we write t when we really mean ct, in the sense that t must be replaced by ct whenever we put dimensions of time, i.e., factors of c, into our formulas. We interpret this as taking $c = 1$ when convenient.

A new twist in formulation (1.30), (1.31) is that the conserved quantities are taken to be the the energy and momentum densities $u = (u^0, u^1) = (T_M^{00}, T_M^{01})$ of the relativistic compressible Euler equations in flat Minkowski spacetime–quantities that, unlike the entries of T, are independent of the metric. Note that, (remarkably), all time derivatives of metric components cancel out from the equations when this change of variables is made, c.f. [10]. We take advantage of this formulation in the numerical method that we introduce here for the study of the initial value problem.

For the proof of Theorem 1, we introduce a fractional step Glimm scheme that employs a Riemann problem step[7] to simulates the source free conservation law $u_t + f(\mathbf{A}, u)_x = 0$, $(\mathbf{A} \equiv Const)$, followed by an ODE step that simulates the effect of the sources present in both f and g, c.f. [16]. Our idea for the numerical scheme is to stagger discontinuities in the metric with discontinuities in the fluid variables so that the conservation law step, as well as the ODE step of the method, are both generated in grid rectangles on which the metric components $\mathbf{A} = (A, B)$, (as well as x), are constant. At the end of each timestep, we solve $A' = h(\mathbf{A}, u, x)$ and re-discretize, to update the metric sources. Part of our proof involves showing that the ODE step $u_t = g(\mathbf{A}, u, x) - \nabla_\mathbf{A} f \cdot \mathbf{A}'$, with h substituted for $\mathbf{A}'$, accounts for both the source term g, as well as the *effective sources* that are due to the discontinuities in the metric components at the boundaries of the grid rectangles.

By our formulation (1.30), (1.31), only the flux f in the conservation law step, depends on $\mathbf{A}$. From this we conclude that the only effect of the metric

---

[7]The Riemann problem is the initial value problem when the initial data is a pair of constant states centered by a jump discontinuity. For a pure conservation law of the form $u_t + f(u)_x = 0$, the solution, which typically only exists for constant states in restricted regions of u-space, consists of elementary waves, c.f. [14, 23].

on the Riemann problem step of the method is to change the wave speeds, but not the states of the waves that solve the Riemann problem. Thus, on the Riemann problem step, when we assume $p = \sigma^2 \rho$, we can apply the estimates obtained in [25], which were originally derived for flat Minkowski spacetime $\mathbf{A} = (1,1)$. Applying these results, it follows that the Riemann problem is *globally* solvable in each grid cell, and the total variation in $\ln \rho$, (the Nishida functional), is non-increasing in time on the Riemann problem step of our fractional step scheme, [25]. Thus we need only estimate the increase in total variation of $\ln \rho$ for the ODE step of the method, in order to obtain a local total variation bound, and hence compactness of the numerical approximations up to some time $T > 0$.

One nice feature of our method is that the ODE that accomplishes the ODE step of the method, turns out to have surprisingly nice properties. Indeed, a phase portrait analysis shows that $\rho > 0$, $|v| < c$ is an invariant region for solution trajectories. (Since x and $\mathbf{A}$ are taken to be constant on the ODE step, the ODE's form an autonomous system at each grid cell.) We also show that even though the ODE's are quadratic in ρ, solutions of the ODE's do not blow up, but in fact remain bounded for all time. It follows that the fractional step scheme is defined and bounded so long as the Courant-Freidrichs-Levy (CFL) condition is maintained, [23]. We show that the CFL bound depends only on the supnorm of the metric component $\|B\|_\infty$, together with the supnorm $\|S\|_\infty$, where $S \equiv S(x,t) = x\rho(x,t)$. We go on to prove that all norms in the problem are bounded by a function that depends only on $\|B\|_\infty$ $\|S\|_\infty$, and $\|TV_L \ln \rho(\cdot,t)\|_\infty$, where the latter denotes the sup of the total variation over intervals of L. By this we show that the solution can be extended up until the first time at which one of these three norms tends to infinity. (Our analysis rules out the possibility that $v \to c$ before one of these norms blows up.) The condition $B \to \infty$ corresponds to the formation of a black hole, and $\rho \to \infty$ corresponds to the formation of a naked singularity, (because the scalar curvature satisfies $R = \{c^2 - 3\sigma^2\}\rho$). It is known that black holes can form in solutions of the Einstein equations, and it is an open problem whether or not naked singularities can form in solutions of the Einstein-Euler equations, or whether we can have $\|S\|_\infty \to \infty$, or $\|TV_L \ln \rho(\cdot,t)\|_\infty \to \infty$, in some other way.

The main technical problem is to prove that the total mass $M_\infty = \frac{\kappa}{2}\int_{r_0}^\infty \rho r^2\, dr$ is bounded. The problem is that, in our estimates, the growth of ρ depends on M and the growth of M depends on ρ, and M is defined by a *non-local* integral. Thus, an error estimate of order Δx for $\Delta \rho$ after one time step, is not sufficient to bound the total mass M_∞ after one time step. [8]

---

[8]The constancy of the total mass reflects the fact that our weak formulation rules our

As a final comment, we note that the total variation bound for system (1.2)-(1.5) is the starting point for the analysis of uniqueness and continous dependence of solutions on the initial data first worked out for homogeneous systems of conservation laws by A. Bressan. This theory, appropriately modified for the source and boundary terms, should also apply to system (1.2)-(1.5). (See [2, 4] and references therein).

1.2   A Locally Inertial Glimm Scheme

We can view this fractional step method as a *locally inertial Glimm scheme* in the sense that it exploits the locally flat character of spacetime. That is, the Riemann Problem step solves the equations $u_t + f(\mathbf{A}, u)_x = 0$ inside grid rectangles $\mathcal{R}_{ij}$. But each grid rectangle defines an "inertial reference frame" because $\mathbf{A} \equiv Const.$ implies the metric is flat in $\mathcal{R}_{ij}$. The boundaries between these local inertial reference frames are the discontinuities that appear along the top, bottom and both sides of the grid rectangles. The term $-\nabla_\mathbf{A} f \cdot \mathbf{A}'$ on the RHS of the ODE step $u_t = g(\mathbf{A}, u, x) - \nabla_\mathbf{A} f \cdot \mathbf{A}'$, accounts for the discontinuities in $\mathbf{A}$ along the sides of the grid rectangles $\mathcal{R}_{ij}$, and the term g in the ODE step, together with the imposition of the constraint $A' = h(\mathbf{A}, u, x)$ at the end of each timestep, account for the discontinuities in $\mathbf{A}$ at the top and bottom of each $\mathcal{R}_{ij}$. It follows that once the convergence of an approximate solution is established, one can just as well replace the true approximate solution by the solution of the Riemann problem in each grid rectangle $\mathcal{R}_{ij}$–the two differ by only order Δx. The resulting appoximation scheme converges to a weak solution of the Einstein equations, and has the property that it solves the compressible Euler equations exactly in local inertial coordinate frames, (grid rectangles), and the transformations between neighboring coordinate frames are accounted for by discontinuities at the coordinate boundaries. In this sense, the fractional step Glimm method is a locally inertial numerical method. It was our search for a locally inertial method that led us to these results, and the success here points to a strategy for obtaining convergent numerical methods in other coordinate systems, c.f. [11].

1.3   The Smoothness Class of the Metric

The RHS of the Einstein equations

---

delta function sources of mass at the shock waves. Note that at points of interaction of shock waves, the Gaussian normal coordinates break down, and so at such points, it is not so easy to analyze the delta function sources from the viewpoint of shock-matching, c.f. [25]

$$G^{ij} = \kappa T^{ij}, \tag{1.41}$$

involve the fluid variables ρ, p and $\mathbf{w}$, thus it follows that when shock waves are present, T is discontinuous, c.f. (1.10). Since the Einstein curvature tensor G on the LHS of (1.41) involves second derivative of the metric g_{ij}, one expects that, in general, the metric components $g_{ij}(x)$ should be at least $C^{1,1}$ functions of the coordinates, (that is, in the smoothness class of continuous functions with Lipschitz continuous first order derivatives), in order that the LHS of (1.41) be free of delta function sources, c.f. [13, 25]. However, the metric components A and B in the solutions of (1.2)-(1.5) constructed here, are only Lipschitz continuous. We know that these solutions are in fact "free of delta function sources" as a consequence of the fact that they are genuine weak solutions of (1.41). It remains an open problem whether or not there exist coordinate transformations that smooth the metric components of these solutions from the smoothness class $C^{0,1}$ up to the class $C^{1,1}$. In such coordinates, (1.10) would hold in the pointwise sense at shock waves, and hence, such a transformation would map weak solutions of the Einstein equations to strong solutions. It was pointed out in [10], (see also [25]), that the transformation that takes an arbitrary spherically symmetric metric over to a metric of form (1.1), necessarily involves derivatives of the metric components, and so the existence of such $C^{1,1}$ coordinates would be consistent with the fact that the A and B that solve (1.2)-(1.5) are only Lipschitz continuous at shock waves. Moreover, in [25, 13] it was shown that for a general smooth shock surface in four dimensional spacetime, such a coordinate transformation always exists, and can be taken to be the Gaussian normal coordinates at the shock surface. But the solutions constructed here can contain arbitrary numbers of interacting shock-waves, of arbitrary strength–and the Gaussian normal coordinate systems break down at points where shock waves interact. With this in mind, we pose the following open question: *Given a weak solution of the Einstein equations for which the metric components are only $C^{0,1}$ functions of the coordinate variables, does there always exist a coordinate transformation that improves the regularity of the metric components to $C^{1,1}$ when the components are viewed as functions of the transformed coordinate variables?* In particular, we ask if this statement is true for the $C^{0,1}$ solutions that we have constructed here?

We believe that this is an interesting question regarding the regularity of solutions of the Einstein equations. Indeed, the Einstein equations are inherently hyperbolic in character; that is, there is finite speed of propagation because all wave speeds are bounded by the speed of light. It follows that, unlike Navier Stokes type parabolic regularizations of the classical compressible Euler equations, incorporating the effects of viscosity and dissipation into Einstein's theory of gravity, cannot alter the fundamental hyperbolic character of the Einstein equations themselves. Thus, even when dissipative effects are accounted for, it is not clear apriori that the corresponding solutions of the Einstein equations will in general be more regular than the solutions that we have constructed here. We also note that the singularity theorems in [12] presume that metrics are in the smoothness class $C^{1,1}$, one degree *smoother* than the solutions we have constructed, c.f. [12], page 284.

In summary, if a transformation exists that impoves the regularity of solutions of the Einstein equations from the class $C^{0,1}$ up to the class $C^{1,1}$, then it defines a mapping that takes weak solutions of the Einstein equations to strong solutions. It then follows that in general relativity, the theory of distributions and the Rankine Hugoniot jump conditions for shock waves need not be imposed on the compressible Euler equations as extra conditions on solutions, but rather must follow as a logical consequence of the strong formulation of the Einstein equations by themselves. If such a transformation does not always exist, then solutions of the Einstein equations are one degree less regular than previously assumed.

2 Preliminaries

The starting point of our analysis is the following theorem, which is a restatement of Theorem 2, [10]. This theorem implies the equivalence of system (1.30),(1.31) with the Einstein equations (1.2)-(1.5) for weak, (shock wave), solutions, so long as (1.3) is treated as a constraint that holds so long as it holds at the boundary $x = r_0$. (We use the variable x in place of r in order to conform with standard notation, c.f. [23]).

Theorem 2 *Let $u(x,t), \mathbf{A}(x,t)$ be weak solutions of (1.30),(1.31) in the domain*

$$D \equiv \{(x,t) : r_0 \leq x < \infty, 0 \leq t < T\}, \qquad (2.1)$$

for some $r_0 > 0$, $T > 0$. Assume that u is in $L^\infty_{loc}(D)$, and that $\mathbf{A}$ is locally Lipschitz continuous in D, by which we mean that for any open ball B centered at a point in D, there is a constant $C > 0$ such that

$$|\mathbf{A}(x_2, t_2) - \mathbf{A}(x_1, t_1)| \leq C\{|x_2 - x_1| + |t_2 - t_1|\}. \qquad (2.2)$$

Then u and $\mathbf{A}$ satisty all four Einstein equations (1.2)- (1.5) throughout D if and only if the equation (1.3),

$$-\frac{\dot{B}}{xB} = \kappa AB T^{01},$$

holds at the boundary $x = r_0$. In this case, it follows that the equivalent forms (1.7), (1.8) of (1.2),(1.3), respectively, also hold in the strong sense throughout D.

Note that for our problem, the constraint (1.8), and therefore (1.3), is implied by the boundary conditions

$$M(r_0, t) = M_{r_0}, \qquad (2.3)$$
$$v(r_0, t) = 0, \qquad (2.4)$$

alone, because, using (1.16), equation (1.8) translates into

$$\dot{M} = -\frac{\kappa}{2}\sqrt{\frac{A}{B}}\frac{c^2 + \sigma^2}{c^2 - v^2} cv\rho x^2,$$

which, in light of (2.3), (2.4), is an identity at the boundary $x = r_0$.

It follows from Theorem 2 that in order to establish Theorem 1, it suffices only to prove the corresponding existence theorem for system (1.30)-(1.31) in domain D. The equation (1.3) will then follow as an identity on weak solutions because it is met at the boundary. It follows that if we construct weak solutions for which v is uniformly bounded and for which ρ decreases fast enough, then we can apply (1.3) as $x \to \infty$ to conclude that

$$\lim_{x\to\infty} \dot{M}(x, t) = 0. \qquad (2.5)$$

This is our strategy for proving that the total mass is finite.

Before stating the main theorem precisely, a few preliminary comments regarding system (1.30)-(1.31) are in order. First note that system (1.30)-(1.31) closes once we express T_M^{11} and T^{22} on the RHS of (1.34), (1.35) and (1.36), as a function of the conserved quantities $u = (u^0, u^1) \equiv (T_M^{00}, T_M^{01})$. From (1.10) it follows that

$$T^{22} = \frac{p}{x^2} = \frac{\sigma^2 \rho}{x^2}, \qquad (2.6)$$

and this can be expressed in terms of u via the mapping (2.21) discussed below. To write T_M^{11} as a function of u, use the identities, (c.f. (4.69),(4,70) of [10]),

$$T_M^{00} - T_M^{11} = \rho c^2 - p \equiv f_1(\rho), \qquad (2.7)$$
$$T_M^{00} T_M^{11} - (T_M^{01})^2 = p\rho c^2 \equiv f_2(\rho). \qquad (2.8)$$

By (2.7),
$$\rho = f_1^{-1}(T_M^{00} - T_M^{11}), \qquad (2.9)$$
and using this in (2.8), one can in general solve (2.8) for T_M^{11}. In the case $p = \sigma^2 \rho$, a calculation gives

$$T_M^{11} = \frac{2\zeta + 1}{2\zeta} \left\{ 1 - \sqrt{1 - \frac{4\zeta}{(2\zeta+1)^2}\left(\zeta + \left[\frac{T_M^{01}}{T_M^{00}}\right]^2\right)} \right\} T_M^{00}, \qquad (2.10)$$

where
$$\zeta = \frac{\sigma^2 c^2}{(c^2 - \sigma^2)^2}. \qquad (2.11)$$

It is readily verified that the quantity under the square root sign is positive so long as
$$\left[\frac{T_M^{01}}{T_M^{00}}\right] < 1 + \frac{1}{2\zeta},$$
which holds in light of (1.20). It follows that (2.10) defines T_M^{11} as a smooth, single valued function of the conserved quantities $(u^0, u^1) \equiv (T_M^{00}, T_M^{01})$. Other than its existence, we will not need the explicit formula for T_M^{11} given in (2.10).

We are free to analyze the state space for system (1.30)-(1.31) in the plane of conserved quantities $u = (u^0, u^1) \equiv (T_M^{00}, T_M^{01})$, in the (ρ, u) plane, or in the plane of Riemann invariants (r, s) which are defined in terms of ρ and v via the special relativistic Euler equations in flat Minkowski spacetime, (assume $p = \sigma^2 \rho$, c.f. [24]),

$$r = \frac{1}{2} \ln \frac{c+v}{c-v} - \frac{K_0}{2} \ln \rho, \qquad (2.12)$$

$$s = \frac{1}{2} \ln \frac{c+v}{c-v} + \frac{K_0}{2} \ln \rho, \qquad (2.13)$$

where
$$K_0 = \frac{2\sigma c}{c^2 + \sigma^2}. \qquad (2.14)$$

(There should be no confusion between "r" the Riemann invariant and "r" the radial coordinate.) It is more convenient for us to use the variables

$$z \equiv s + r = K_0 \ln \rho, \tag{2.15}$$
$$w \equiv s - r = \ln \frac{c+v}{c-v}, \tag{2.16}$$

and we let $\mathbf{z}$ denote the vector

$$\mathbf{z} = (z, w) \equiv \left(K_0 \ln \rho, \ln \frac{c-v}{c+v}\right). \tag{2.17}$$

Given this, we use the following notation: As usual, the double norm $\|\cdot\|$ applied to a vector denotes Euclidean norm, so e.g., $\|u\| \equiv \sqrt{(u^0)^2 + (u^1)^2}$ and $\|\mathbf{z}\| \equiv \sqrt{(z)^2 + (w)^2}$, and the single norm $|\cdot|$, when applied to scalars, denotes the the regular absolute value. But we use the special notation that $|\cdot|$, when applied to a vector, denotes the change in the z-component across the vector, so that, e.g.,

$$|\mathbf{z}| \equiv |z|. \tag{2.18}$$

Similarly, if γ denotes a wave with left state $\mathbf{z}_L$ and right state $\mathbf{z}_L$, (see (3.10) and (4.13)-(4.15) below), then we let

$$\|\gamma\| \equiv \sqrt{|z_R - z_L|^2 + |w_R - w_R|^2}, \tag{2.19}$$
$$|\gamma| \equiv |z_R - z_L|, \tag{2.20}$$

and we refer to $|\gamma|$ as the strength of the wave γ, c.f. [24, 16].

Equations (1.15), (1.16), and (2.12)-(2.16), define the mappings $\Psi : (\rho, v) \to (u^0, u^1)$ and $\Phi : (\rho, v) \to (z, w)$,

$$\begin{pmatrix} u^0 \\ u^1 \end{pmatrix} = \Psi \begin{pmatrix} \rho \\ v \end{pmatrix} \equiv \begin{pmatrix} \frac{c^4 + \sigma^2 v^2}{c^2 - v^2} \rho c^2 \\ \frac{(c^2 + \sigma^2) cv}{c^2 - v^2} \rho \end{pmatrix}, \tag{2.21}$$

$$\begin{pmatrix} z \\ w \end{pmatrix} = \Phi \begin{pmatrix} \rho \\ v \end{pmatrix} \equiv \begin{pmatrix} K_0 \ln \rho \\ \ln \frac{c+v}{c-v} \end{pmatrix}. \tag{2.22}$$

The following proposition states that the mappings Ψ and Φ define one to one regular maps between the respective domains:

Proposition 1 *The mapping*

$$\Phi : D \to R \tag{2.23}$$

defined by (2.21) is smooth, one-to-one and onto, from domain

$$\mathcal{D} = \{(\rho, v) : 0 < \rho < \infty, |v| < c\}, \tag{2.24}$$

to range

$$R = \{(u^0, u^1) : 0 < u^0 < \infty, |u^1| < \infty\}; \tag{2.25}$$

and the mapping

$$\Phi : \mathcal{D} \to \hat{R}, \tag{2.26}$$

defined in (2.22), is smooth, one-to-one and onto from domain $\mathcal{D}$ to

$$\hat{R} = \{(z, w) : -\infty < z < +\infty, -\infty < w < +\infty\}. \tag{2.27}$$

Proof: This follows directly from (2.21) and (2.22).

The goal of this paper is to prove the following theorem:

Theorem 3 *Let $u_0(x) \equiv (u_0^0(x), u_0^1(x)) = \Psi(\rho_0(x), v_0(x)) = \Psi \circ \Phi^{-1}(z_0(x), w_0(x))$ and $\mathbf{A}_0(x) = (A_0(x), B_0(x))$ denote initial data for system (1.30),(1.31), defined for $x \geq r_0$. Assume that there exists positive constants V, L, and $\bar{v}$, such that*

$$TV_{[x,x+L]} \ln \rho_0(\cdot) < V, \tag{2.28}$$

$$TV_{[x,x+L]} \ln \tfrac{c+v_0(\cdot)}{c-v_0(\cdot)} < V, \tag{2.29}$$

$$|v_0(x)| < \bar{v}, \tag{2.30}$$

for all $x \geq r_0$. Assume that $B_0(x) = \frac{1}{1 - \frac{2M_0(x)}{x}}$, where the initial mass function $M_0(x)$ is given by

$$M_0(x) = M_{r_0} + \frac{\kappa}{2} \int_{r_0}^{x} u_0^0(r) r^2 \, dr, \tag{2.31}$$

(c.f. (1.18)), and assume that M_0 satisfies the conditions

$$\lim_{x \to \infty} M_0(x) = M_\infty < \infty, \tag{2.32}$$

and

$$1 - \frac{2M_0(x)}{x} = B_0^{-1}(x) > \bar{B}^{-1} > 0, \tag{2.33}$$

respectively, for some fixed positive constants $M_{r_0} < M_\infty$, and $\bar{B} < \infty$. Assume finally that

$$A_0(x) = A_{r_0} \exp \int_{r_0}^{x} \left\{ \frac{B_0(r) - 1}{r} + \kappa r B_0(r) T_M^{11}(u_0(r)) \right\} dr \qquad (2.34)$$

for some fixed positive constant $A_{r_0} > 0$, so that

$$A_0(r_0) = A_{r_0} > 0. \qquad (2.35)$$

Given this, we conclude that there exists $T > 0$, and functions $u(x,t), \mathbf{A}(x,t)$ defined on $x \geq r_0$, $0 \leq t < T$, such that $u(x,t), \mathbf{A}(x,t)$ is a weak solution of system (1.26),(1.27), (1.2)-(1.4), together with the initial-boundary conditions

$$\rho(x,0) = \rho_0(x), \quad v(x,0) = v_0(x), \qquad (2.36)$$

$$\mathbf{A}(r_0, t) = \left(A_{r_0}, \frac{1}{1 - \frac{2Mr_0}{r_0}} \right), \qquad (2.37)$$

$$v(r_0, t) = 0. \qquad (2.38)$$

Moreover, the solution $u, \mathbf{A}$ satisfies the following:
(i) For each $t \in [0,T)$ there exists a constant $V(t) < \infty$ such that

$$TV_{[x,x+L]} \ln \rho(\cdot, t') < V(t), \qquad (2.39)$$

$$TV_{[x,x+L]} \ln \frac{c + v(\cdot, t')}{c - v(\cdot, t')} < V(t), \qquad (2.40)$$

for all $t' \leq t$.
(ii) For each $x \geq r_0$ and $t \in [0,T)$,

$$0 < A(x,t), B(x,t) < \infty, \qquad (2.41)$$

and

$$\lim_{x \to \infty} M(x,t) = M_\infty. \qquad (2.42)$$

(iii) For each closed bounded set $\mathcal{U} \subset \{(x,t) : x \geq r_0, \ 0 \leq t < T\}$, there exists a constant $C(\mathcal{U}) < \infty$ such that,

$$\|\mathbf{A}(x_2, t_2) - \mathbf{A}(x_1, t_1)\| < C(\mathcal{U}) \left\{ |x_2 - x_1| + |t_2 - t_1| \right\}, \qquad (2.43)$$

and

$$\int_{r_0}^{x} \|u(r, t_2) - u(r, t_1)\| dr < C(\mathcal{U}) |t_2 - t_1|. \qquad (2.44)$$

Here, (2.39) and (2.40) imply that the functions $z(\cdot,t)$ and $w(\cdot,t)$ are functions of locally bounded total variation at each fixed time $t < T$, and the bounds are uniform over bounded sets in $x \geq r_0$, $0 \leq t < T$. Estimates (2.39) and (2.40) also imply that $\rho > 0$ and $|v| < c$, and therefore that $u^0 > 0$ throughout $x \geq r_0$, $0 \leq t < T$. The inequality (2.41) says that $B = \frac{1}{1-\frac{2M}{x}} > 0$, and hence that $\frac{2M}{x} < 1$ for $t < T$, the condition that no black holes have formed before time T. Inequality (2.43) says that the metric components A and B are locally Lipschitz continuous functions in $x \geq r_0$, $0 \leq t < T$, and (2.43) says that $u(x,t)$ is L^1-Lipschitz continuous in time, uniformly on bounded sets. Note that (2.31), (2.34) are included to guarantee that equations (1.2) and (1.4), (and so also (1.31)), are satisfied at time $t = 0$.

3 The Fractional Step Scheme.

In this section we define the approximate solutions $u_{\Delta x}$, $\mathbf{A}_{\Delta x} = (A_{\Delta x}, B_{\Delta x})$ of system (1.30), (1.31) constructed by a fractional step Glimm scheme. Again, we have set $x \equiv x^1 \equiv r$, and we write t in place of ct, in the sense that t must be replaced by ct whenever we put dimensions of time, (that is, factors of c), into our formulas.

Let $\Delta x << 1$ denote a mesh length for space and Δt a mesh length for time, and assume that

$$\frac{\Delta x}{\Delta t} = \Lambda, \tag{3.1}$$

so that Λ^{-1} is the Courant number. We choose

$$\Lambda \geq Max\left\{2\sqrt{\frac{A}{B}}\right\}, \tag{3.2}$$

where the maximum is taken over all values that appear in the approximate solution. This guarantees the Courant-Friedrichs-Levy (CFL) condition, the condition that the mesh speed be greater than the maximum wave speed in the problem. (That is, $\sqrt{\frac{A}{B}}$ is the speed of light in Schwarzschild coordinates, and the factor of two accounts for the fact that waves emanate from the center of the mesh rectangles in our approximation scheme. Of course, as part of our proof, we must show that the maximum on the RHS of (3.2) exists.) Let (x_i, t_j) be mesh points in an unstaggered grid defined on the domain

$$D = \{r_0 \leq x \leq \infty, t \geq 0\}, \tag{3.3}$$

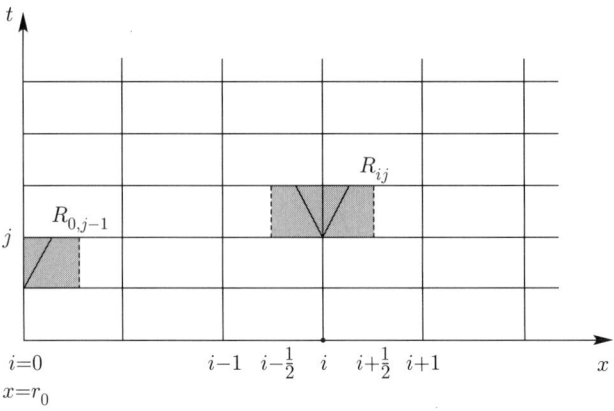

Figure 1: The mesh rectangles $\mathcal{R}_{ij}$

by setting

$$x_i = r_0 + i\Delta r,$$
$$t_j = j\Delta t,$$

c.f. Figure 1.

Each mesh point (x_i, t_j), $i \geq 0$, $j \geq 0$, is positioned at the bottom center of the grid rectangle $\mathcal{R}_{ij}$,

$$\mathcal{R}_{ij} = \{x_{i-\frac{1}{2}} \leq x < x_{i+\frac{1}{2}}, \ t_j \leq t < t_{j+1}\}, \tag{3.4}$$

where $x_{i+\frac{1}{2}} = (i+\frac{1}{2})\Delta x$. Let $\mathcal{R}_{i_0 j}$ denote the half rectangle $\{x_{i_0} \leq x < x_{i_0+\frac{1}{2}}, \ t_j \leq t < t_{j+1}\}$ at the boundary $x = r_0$. In the approximation scheme, the metric source $\mathbf{A} = (A, B)$ is approximated by the constant value $\mathbf{A}_{ij}$ in each grid rectangle $\mathcal{R}_{ij}$, so set

$$\mathbf{A}_{\Delta x}(x, t) = \mathbf{A}_{ij} \text{ for } (x, t) \in \mathcal{R}_{ij}, \tag{3.5}$$

for values of $\mathbf{A}_{ij}$ to be defined presently. It follows that $\mathbf{A}_{\Delta x}$ is discontinuous along each line $x = x_{i+\frac{1}{2}}$, $i = 0, ..., \infty$, and at each time $t = t_j$. In our definition below, values of $\mathbf{A}_{ij}$ are determined from values $\mathbf{A}_{i,j-1}$ and $u_{\Delta x}$ at time t_j-, by solving (1.31), using the boundary condition $\mathbf{A} = \mathbf{A}_{r_0} = \left(A_{r_0}, \frac{1}{1-\frac{2Mr_0}{r_0}} \right)$ at the boundary $x = r_0$.

We now define $u_{\Delta x}$ by induction. First assume that $u_{\Delta x}$ is given by piecewise constant states u_{ij} at time $t = t_j+$ as follows:

$$u_{\Delta x}(x, t) = u_{ij} \text{ for } x_i \leq x < x_{i+1},\ t = t_j+. \tag{3.6}$$

This poses the Riemann problem

$$u_0(x) = \begin{cases} u_L = u_{i-1,j} & x < x_i, \\ u_R = u_{ij} & x > x_i, \end{cases} \tag{3.7}$$

for the system

$$u_t + f(\mathbf{A}_{ij}, u)_x = 0, \tag{3.8}$$

at the bottom center of each mesh rectangle $\mathcal{R}_{ij}$, $i \geq 1$. When $i = 0$, the boundary condition $v = 0$ at $x_0 = r_0$ replaces the left state, and so in this case, the piecewise constant state u_{0j} at time $t = t_j+$ poses the boundary Riemann problem

$$u_0(x) = \begin{cases} v = 0 & x = r_0, \\ u_R = u_{0,j} & x > r_0, \end{cases} \tag{3.9}$$

Let $u_{ij}^{RP}(x, t)$ denote the solution of (3.6), (3.7) for $(x, t) \in \mathcal{R}_{ij}$, and let

$$u_{\Delta x}^{RP}(x, t) = u_{ij}^{RP}(x, t) \text{ for } (x, t) \in \mathcal{R}_{ij}. \tag{3.10}$$

Equation (3.10) defines the Riemann problem step of the fractional step scheme. Note that since $\mathbf{A}_{\Delta x} = \mathbf{A}_{ij}$ is constant in each $\mathcal{R}_{ij}$, it follows that system (3.8) is just the special relativistic Euler equations for $p = \sigma^2 \rho$, with a rescaled flux. We discuss the solution of this Riemann problem in detail in Section 4. We conclude there that the solution $u_{ij}^{RP}(x, t)$ consists of a 1-wave γ_{ij}^1 followed by a 2-wave γ_{ij}^2 for all $i > 0$, it consists of a single 2-wave $\gamma_{0j}^2 = 0$ at the boundary $i = 0$, and the waves γ_{ij}^p all have sub-luminous speeds so long as (3.2) holds. It follows that (3.2) guarantees that the waves in the Riemann problem (3.7), (3.7), never leave $\mathcal{R}_{ij}$ in one time step, c.f. Proposition 3 below.

The Riemann problem step of the method ignores the effect of the source term g in system (1.30), and also ignores the effect of the discontinuities in the flux $f(\mathbf{A}, u)$ due to discontinuities in $\mathbf{A}$ at the boundaries $x_{i-\frac{1}{2}}$ of $\mathcal{R}_{ij}$. These effects are accounted for in the ODE step. For the ODE step of the fractional step scheme, the idea is to use the Riemann problem solutions as initial data, and solve the ODE's

$$u_t = G(\mathbf{A}, u, x) \equiv g - \mathbf{A}' \cdot \nabla_{\mathbf{A}} f, \tag{3.11}$$

for one time step, thus defining the approximate solution in $\mathcal{R}_{ij}$. The first term on the RHS of (3.11) accounts for the sources on the RHS of (1.30), and the second term accounts for the discontinuities in $\mathbf{A}$ at the boundaries $x_{i-\frac{1}{2}}$. Now by (1.38),

$$g \equiv g(\mathbf{A}, u, x) = \left(g^0(\mathbf{A}, u, x), g^1(\mathbf{A}, u, x)\right),$$

where

$$g^0(\mathbf{A}, u, x) = -\frac{2}{x}\sqrt{\frac{A}{B}} T_M^{01}, \tag{3.12}$$

$$g^1(\mathbf{A}, u, x) = -\frac{1}{2}\sqrt{\frac{A}{B}} \left\{ \frac{4}{x} T_M^{11} + \frac{(B-1)}{x}(T_M^{00} - T_M^{11}) \right. \tag{3.13}$$

$$\left. + 2\kappa x B(T_M^{00} T_M^{11} - (T_M^{01})^2) - 4x T^{22} \right\}.$$

By (1.34),

$$\nabla_{\mathbf{A}} f \equiv \nabla_{\mathbf{A}} f(\mathbf{A}, u) \equiv \left(\nabla_{\mathbf{A}} f^0, \nabla_{\mathbf{A}} f^1\right) = \left(\frac{1}{2}\frac{1}{\sqrt{AB}} T_M^{01}, -\frac{1}{2}\frac{1}{B\sqrt{AB}} T_M^{11}\right), \tag{3.14}$$

and by (1.40),

$$\mathbf{A}' = h \equiv \left(h^0(\mathbf{A}, u, x)), h^1(\mathbf{A}, u, x))\right), \tag{3.15}$$

where

$$h^0(\mathbf{A}, u, x) = \frac{(B-1)A}{x} + \kappa x A B T_M^{11},$$
$$h^1(\mathbf{A}, u, x) = -\frac{(B-1)B}{x} + \kappa x B^2 T_M^{00}. \tag{3.16}$$

It follows from (3.14)-(3.16) that

$$\mathbf{A}' \cdot \nabla_{\mathbf{A}} f(\mathbf{A}, u, x) = \frac{1}{2}\sqrt{\frac{A}{B}} \delta\left(T_M^{01}, T_M^{11}\right), \tag{3.17}$$

where

$$\delta = \frac{A'}{A} - \frac{B'}{B} = \frac{2(B-1)}{x} - \kappa x B \left(T_M^{00} - T_M^{11}\right). \quad (3.18)$$

Using (3.12), (3.13) and (3.18) and simplfying, we find that the ODE step should be

$$u_t = G, \quad (3.19)$$

where

$$G^0(\mathbf{A}, u, x) = -\frac{1}{2}\sqrt{\frac{A}{B}} T_M^{01} \left\{\frac{2(B+1)}{x} - \kappa x B \left(T_M^{00} - T_M^{11}\right)\right\}, \quad (3.20)$$

$$G^1(\mathbf{A}, u, x) = -\frac{1}{2}\sqrt{\frac{A}{B}} \left\{\frac{4}{x}T_M^{11} + \frac{B-1}{x}\left(T_M^{00} + T_M^{11}\right) + \quad (3.21)\right.$$
$$\left. \kappa x B \left[T_M^{00} T_M^{11} - 2\left(T_M^{01}\right)^2 + \left(T_M^{11}\right)^2\right] - 4xT^{22}\right\}.$$

Since $u = (T_M^{00}, T_M^{01})$, and T_M^{11}, T^{22} are given as functions of u in (2.6), (2.10), respectively, it follows that the right hand sides of (3.20) and (3.21) determine well defined functions of $(\mathbf{A}, u, x)$. It follows that G, as defined in (3.20), (3.21) also satisfies

$$G(\mathbf{A}, u, x) = g(\mathbf{A}, u, x) - \mathbf{A}' \cdot \nabla_{\mathbf{A}} f(\mathbf{A}, u, x),$$

where (3.12), (3.13) and (3.17) define g and $\mathbf{A}' \cdot f$ as functions of $(\mathbf{A}, u, x)$.

We can now define the ODE step of the method. Let $\hat{u}(t, u_0)$ denote the solution to the initial value problem

$$\hat{u}_t = G(\mathbf{A}_{ij}, \hat{u}, x) = g(\mathbf{A}_{ij}, \hat{u}, x) - \mathbf{A}' \cdot \nabla_{\mathbf{A}} f(\mathbf{A}_{ij}, \hat{u}, x),$$
$$\hat{u}(0) = u_0, \quad (3.22)$$

where $G(\mathbf{A}, \hat{u}, x)$ is defined in (3.20), (3.21), and $g(\mathbf{A}, u, x)$ and $\mathbf{A}' \cdot f(\mathbf{A}, u, x)$ are defined in (3.12), (3.13) and (3.17), respectively. It follows that

$$\hat{u}(t, u_0) - u_0 = \int_0^t \hat{u}_t \, dt$$
$$= \int_0^t \{g(\mathbf{A}_{ij}, \hat{u}(\xi, u_0), x) - \mathbf{A}' \cdot \nabla_{\mathbf{A}} f(\mathbf{A}_{ij}, \hat{u}(\xi, u_0), x)\} \, d\xi.$$

Define the approximate solution $u_{\Delta x}(x,t)$ on each mesh rectangle $\mathcal{R}_{ij}$ by the formula

$$u_{\Delta x}(x,t) = u_{\Delta x}^{RP}(x,t) + \int_{t_j}^{t} \left\{ G(\mathbf{A}_{ij}, \hat{u}(\xi - t_j, u_{\Delta x}^{RP}(x,t)), x) \right\} d\xi \quad (3.23)$$

$$= u_{\Delta x}^{RP}(x,t) + \int_{t_j}^{t} \left\{ g(\mathbf{A}_{ij}, \hat{u}(\xi - t_j, u_{\Delta x}^{RP}(x,t)), x) \right\} d\xi$$

$$- \int_{t_j}^{t} \left\{ \mathbf{A}' \cdot \nabla_{\mathbf{A}} f(\mathbf{A}_{ij}, \hat{u}(\xi - t_j, u_{\Delta x}^{RP}(x,t)), x) \right\} d\xi.$$

Thus on each mesh rectangle $\mathcal{R}_{ij}$, $u_{\Delta x}(x,t)$ is equal to $u_{\Delta x}^{RP}(x,t)$ plus a correction that defines the ODE step of the method.

To complete the definition of $u_{\Delta x}$ by induction, it remains only to define the constant states $A_{i,j+1}$ on $\mathcal{R}_{i,j+1}$, and $u_{i,j+1} = u_{\Delta x}(x, t_{j+1}+)$ for $x_i \le x < x_{i+1}$, in terms of the values of $u_{\Delta x}, \mathbf{A}_{\Delta x}$ defined for $t_j \le t < t_{j+1}$. For this we use Glimm's method of random choice, c.f. [8, 23]. Thus let

$$\mathbf{a} \equiv \{a_j\}_{j=0}^{\infty} \in \mathbf{\Pi}, \quad (3.24)$$

denote a (fixed) random sequence, $0 < a_j < 1$, where $\mathbf{\Pi}$ denotes the infinite product measure space $\Pi_{i=0}^{\infty}(0,1)_j$, where $(0,1)_j$ denotes the unit interval $(0,1)$ endowed with Lebesgue measure, $0 < j < \infty$. (For convenience, assume WLOG that $a_0 = \frac{1}{2}$.) Then, assuming that $u_{\Delta x}, \mathbf{A}_{\Delta x}$ is defined up to time $t < t_{j+1}$, define

$$u_{i,j+1} = u_{\Delta x}(x_i + a_{j+1}\Delta x, t_{j+1}-), \quad (3.25)$$

$$M_{\Delta x}(x, t_{j+1}) = M_{r_0} + \frac{\kappa}{2} \int_{r_0}^{x} u_{\Delta x}^0(r, t_{j+1}-)r^2 \, dr, \quad (3.26)$$

c.f. (1.18).[9] In terms of these, define the functions

$$B_{\Delta x}(x, t_{j+1}) = \frac{1}{1 - \frac{2M_{\Delta x}(x, t_{j+1})}{x}}, \quad (3.27)$$

and

$$A(x, t_{j+1}) = A_{r_0} \exp \int_{r_0}^{x} \left\{ \frac{B_{\Delta x}(r, t_{j+1}) - 1}{r} + \kappa r B_{\Delta x}(r, t_{j+1}) T_M^{11}(u_{\Delta x}(r, t_{j+1})) \right\} dr, \quad (3.28)$$

---

[9] By (3.25), the approximate solution depends on the choice of sample sequence $\mathbf{a}$. In the last section, we prove that for almost every choice of sample sequence, a subsequence of approximate solutions converges to a weak solution of (1.31).

c.f. (1.6) and (1.4). Finally, in terms of these, define

$$M_{i,j+1} = M(x_i, t_{j+1}), \tag{3.29}$$

$$B_{i,j+1} = B(x_i, t_{j+1}) = \frac{1}{1 - \frac{2M_{i,j+1}}{x_i}}, \tag{3.30}$$

and

$$A_{i,j+1} = A(x_i, t_{j+1}). \tag{3.31}$$

Let $\mathbf{A}_{i,j+1} = (A_{i,j+1}, B_{i,j+1})$ denote the constant value for $A_{\Delta x}$ on $\mathcal{R}_{i,j+1}$. This completes the definition of the approximate solution $u_{\Delta x}$ by induction. Note that (3.26)-(3.28) imply that when $\rho > 0$, $|v| < c$, we have

$$B_{\Delta x}(x, t_j) \geq 1, \tag{3.32}$$

$$B_{\Delta x}(r_0, t_j) = \frac{1}{1 - \frac{2Mr_0}{r_0}} \equiv B_{r_0}, \tag{3.33}$$

$$A_{\Delta x}(x, t_j) \geq A_{r_0}, \tag{3.34}$$

for all $x \geq r_0, j \geq 0$. Note also that as a consequence of (3.26), (3.27) and (3.28), equations (1.2) and (1.4) hold in the form

$$\frac{B'_{\Delta x}(x,t_j)}{B} = -\frac{B_{\Delta x}(x,t_j) - 1}{x} + \kappa B_{\Delta x}(x, t_j) x T_M^{11}(u_{\Delta x}(x, t_j)), \tag{3.35}$$

$$\frac{A'_{\Delta x}(x,t_j)}{A} = +\frac{B_{\Delta x}(x,t_j) - 1}{x} + \kappa B_{\Delta x}(x, t_j) x T_M^{00}(u_{\Delta x}(x, t_j)). \tag{3.36}$$

Therefore,

$$\frac{\partial}{\partial x} \ln \{A_{\Delta x}(x, t_j) B_{\Delta x}(x, t_j)\} = \frac{A'}{A} + \frac{B'}{B}$$
$$\leq 4\kappa x B_{\Delta x}(x, t_j)(T_M^{00}(u_{\Delta x}(x, t_j)) + T_M^{11}(u_{\Delta x}(x, t_j))).$$

Integrating this from r_0 to x yields

$$A_{\Delta x}(x, t_j) B_{\Delta x}(x, t_j) \leq A_{r_0} B_{r_0} \exp\left\{\frac{8}{r_0} \int_{r_0}^{x} B_{\Delta x}(x, t_j) \frac{\kappa}{2} r^2 T_M^{00}(u_{\Delta x}(x, t_j))\right\}. \tag{3.37}$$

Inequalities (3.35)-(3.37) directly imply the following proposition:

Proposition 2 *Assume that there exist positive constants $\bar{M}$, $\bar{B}$, $\bar{S}$, $\bar{v}$, and integer $J > 0$, such that the approximate solution $u_{\Delta x}$, $\mathbf{A}_{\Delta x}$, defined as above, exists and satisfies*

$$M_{\Delta x}(x, t_j) \leq \bar{M}, \tag{3.38}$$

$$B_{\Delta x}(x, t_j) \leq \bar{B}, \tag{3.39}$$

$$0 \leq S_{\Delta x}(x, t_j) \equiv |x \rho_{\Delta x}(x, t_j)| \leq \bar{S} < \infty, \tag{3.40}$$

and

$$|v_{\Delta x}(x, t_j)| \leq \bar{v} < c, \tag{3.41}$$

for all $x \geq r_0$, $j \leq J$, so that by (1.15),

$$0 \leq x u^0_{\Delta x}(x, t_j) \leq \frac{c^2 + \sigma^2 \bar{v}^2}{c^2 - \bar{v}^2} \bar{S}. \tag{3.42}$$

Then

$$0 < \frac{A_{r_0}}{B_{\Delta x}(x, t_j)} \leq \frac{A_{\Delta x}(x, t_j)}{B_{\Delta x}(x, t_j)} \leq A_{\Delta x}(x, t_j) \leq A_{\Delta x}(x, t_j) B_{\Delta x}(x, t_j)$$
$$\leq A_{r_0} B_{r_0} exp\left\{\frac{8\bar{B}\bar{M}}{r_0}\right\} \equiv G_{AB}(\bar{B}, \bar{M}), \tag{3.43}$$

and

$$|A'_{\Delta x}(x, t_j)| \leq \left(\frac{1}{r_0} + \kappa \frac{c^2 + \sigma^2 \bar{v}^2}{c^2 - \bar{v}^2} \bar{S}\right) G_{AB}(\bar{B}, \bar{M}), \tag{3.44}$$

$$|B'_{\Delta x}(x, t_j)| \leq \left(\frac{1}{r_0} + \kappa \frac{c^2 + \sigma^2 \bar{v}^2}{c^2 - \bar{v}^2} \bar{S}\right) \bar{B}^2 \tag{3.45}$$

for all $x \geq r_0$, and $j \leq J$.

Note that by (3.30), (3.31), (3.43)-(3.45) apply with $A_{\Delta x}(x, t_j)$, $B_{\Delta x}(x, t_j)$, replaced by A_{ij}, B_{ij}, respectively. Note also that (3.43) implies that

$$\Lambda = 2\sqrt{G_{AB}} \tag{3.46}$$

suffices to guarantee the CFL condition (3.2), and note that (3.44) and (3.45) imply

$$\left\| \frac{\Delta \mathbf{A}_{\Delta x}}{\Delta x} \right\| \leq \left(\frac{1}{r_0} + \kappa \frac{c^2 + \sigma^2 \bar{v}^2}{c^2 - \bar{v}^2} \bar{S} \right) (\bar{B}^2 + G_{AB}(\bar{B}, \bar{M})), \tag{3.47}$$

where

$$\frac{\Delta \mathbf{A}_{\Delta x}}{\Delta x} = \frac{\mathbf{A}_{i+1,j} - \mathbf{A}_{ij}}{\Delta x}, \tag{3.48}$$

which gives the Lipschitz continuity in x of $A_{\Delta x}$ and $B_{\Delta x}$, respectively.

Proof: Inequality (3.43) follows directly form (3.37) in light of (7.30) and (3.26), and (3.44), (3.45) follow directly from (3.35), (3.36), and (3.43).

4 The Riemann Problem Step

In this section we discuss u_{ij}^{RP}, the solutions which constitute the Riemann problem step in the construction of $u_{\Delta x}$. For fixed (i, j), $u_{ij}^{RP}(x, t)$ is defined in (3.7), (3.8) as the solution of the Riemann problem

$$u_t + f(\mathbf{A}_{ij}, u)_x = 0, \tag{4.1}$$

$$u_0(x) = \left\{ \begin{array}{ll} u_L = u_{i-1,j} & x < 0 \\ u_R = u_{ij} & x \geq 0 \end{array} \right\}, \tag{4.2}$$

with the origin translated to the bottom center (x_i, t_j) of the mesh rectangle $\mathcal{R}_{ij} \equiv \{(x, t) : x_{i-\frac{1}{2}} < x \leq x_{i+\frac{1}{2}}, t_j \leq t < t_{j+1}\}$. Vector $\mathbf{A}_{ij}$ is constant on $\mathcal{R}_{ij}$. Assuming $p = \sigma^2 \rho$, system (4.1) takes the form

$$(T_M^{00})_{,t} + \left(\sqrt{\frac{A_{ij}}{B_{ij}}} T_M^{01} \right)_{,x} = 0, \tag{4.3}$$

$$(T_M^{01})_{,t} + \left(\sqrt{\frac{A_{ij}}{B_{ij}}} T_M^{11} \right)_{,x} = 0, \tag{4.4}$$

where T_M^{11} is given as a function of T_M^{00} and T_M^{01} in (2.10).

Proposition 3 *Assume that u_L and u_R correspond to values of ρ and v that lie in the region $\rho > 0$, $-c < v < c$. Then the Riemann problem (4.1), (4.2) has a unique solution consisting of elementary waves: shock waves and rarefaction waves. The solution is scale invariant, (is a function of x/t), and consists of a 1-wave γ_{ij}^1 followed by a 2-wave γ_{ij}^2. Moreover, the CFL condition (3.2) guarantees that the speeds of the waves are always smaller than the mesh speed $\frac{\Delta x}{\Delta t} = Max\left\{2\sqrt{\frac{A}{B}}\right\}$, and thus waves never interact during one time step.*

Proof: System (4.3)-(4.4) is the relativistic compressible Euler equations $div T_M = 0$ in flat Minkowski spacetime, except for the constant factor $\sqrt{\frac{A_{ij}}{B_{ij}}}$ that multiplies the flux. Now the factor $\sqrt{\frac{A_{ij}}{B_{ij}}}$ changes the speeds of the waves, but does not affect the values of u on the elementary waves γ_{ij}^p. Indeed, the scale change $\bar{t} \to t/\sqrt{A_{ij}/B_{ij}}$ converts (4.1) into the Minkowski space problem $div T_M = 0$, and so it follows from the frame invariance of the compressible Euler equations that (s, u_L, u_R) satisfies the Rankine-Hugoniot jump conditions

$$s[u] = [f] = \sqrt{\frac{A_{ij}}{B_{ij}}}[f_M], \qquad (4.5)$$

for system (4.1), if and only if $(\bar{s}, u)$ satisfies the Minkowski jump conditions

$$\bar{s}[u] = [f_M], \qquad (4.6)$$

where

$$s = \sqrt{\frac{A_{ij}}{B_{ij}}}\bar{s}. \qquad (4.7)$$

(Recall that a shock with left state u_L, right state u_R, and speed s, is a weak solution of a conservation law $u_t + f(u)_x = 0$ if and only if the Rankine-Hugoniot jump relations $s[u] = [f]$ are satisfied.) Here f denotes the flux in (4.1), $f_M = f(1,1,u)$ denotes the standard Minkowski flux, and $[\cdot]$ denotes the jump in a quantity from left to right across a shock. Thus the i-shock curves for system (4.1) agree with the i-shock curves for the system $u_t + f_M(u)_x = 0$, when $\mathbf{A}_{ij} = (A_{ij}, B_{ij}) = (1,1)$, [23]. Moreover, since $[u]$ tends to an eigen-direction and s tends to an eigenspeed as $[u] \to 0$ across a shock, it follows that the i-rarefaction curves $\mathbf{R}_i$ and i-shock curves $\mathbf{S}_i$ for system (4.1) are the same as the curves for the Minkowski system $u_t + f_M(u)_x = 0$, c.f. [23, 24, 8, 14, 3]. It follows that the factor $\sqrt{\frac{A_{ij}}{B_{ij}}}$ changes the speeds of the waves, but does not affect the values of u on the elementary waves γ_{ij}^p, as claimed.

It was shown in [24] that the Riemann problem for system $u_t + f_M(u)_x = 0$ has a unique solution consisting of a 1-wave followed by a 2-wave, and all wave speeds are subluminous so long as $\rho > 0$, $-c < v < c$. If we denote this solution by $[u_L, u_R]_M(x, t)$, then, (assuming $\rho > 0$, $-c < v < c$), it follows from (4.7) that the solution of (4.1), (4.2) is given by $[u_L, u_R](x, t) = [u_L, u_R]_M(x, \sqrt{\frac{A_{ij}}{B_{ij}}} t)$. Since, by [24], all shock and characteristic speeds are sub-luminous for the Minkowski problem $\text{div} T_M = 0$, $p = \sigma^2 \rho$, it follows from (4.7) that wave speeds in the solution of the Riemann problem (4.1), (4.2) are bounded by $\sqrt{\frac{A}{B}}$, the speed of light in standard Schwarzschild coordinates. This verifies that if $\rho > 0$, $-c < v < c$, then the CFL condition (3.2) guarantees that all wave speeds in the solution u_{ij}^{RP} are bounded by the mesh speed $\frac{\Delta x}{\Delta t} = \Lambda$. Note that this implies that the constant states $u_{i-1,j}$, u_{ij} are maintained along the left and right boundaries of $\mathcal{R}_{ij}$ in the approximate solution u_{ij}^{RP}. $\square$

For fixed $\mathbf{A_{ij}} = (A_{ij}, B_{ij})$, let

$$[u_L, u_R] \equiv [u_L, u_R](x, t), \tag{4.8}$$

denote the solution of the Riemann problem (4.1), (4.2), and write

$$[u_L, u_R] = (\gamma^1, \gamma^2), \tag{4.9}$$

to indicate that the solution $[u_L, u_R](x, t)$ consists of the 1-wave γ^1 followed by the 2-wave γ^2. An elementary wave γ is itself a solution of a Riemann problem, in which case we write $[u_L, u_R] = \gamma$, and we call u_L and u_R the right and left states of the wave γ, respectively. In this case, define $|\gamma|$, the strength of the wave γ, by

$$|\gamma| = |K_0 \ln(u_L) - K_0 \ln(u_R)| = \left| K_0 \ln\left(\frac{u_L}{u_R}\right) \right|, \tag{4.10}$$

c.f. (2.14). For the general case $[u_L, u_R] = (\gamma^1, \gamma^2)$, we define the strength of the Riemann problem as the sum of the strengths of its elementary waves,

$$|[u_L, u_R]| = |\gamma^1| + |\gamma^2|. \tag{4.11}$$

The following proposition, special to the case $p = \sigma^2 \rho$, states that the sum of the strengths of elementary waves are non-increasing during wave interactions, so long as $\mathbf{A}_{ij}$ is constant. (This property was first identified by Nishida in non-relativistic case, [20].)

Proposition 4 *Assume that $\mathbf{A}_{ij}$ is fixed. Let u_L, u_M, and u_R be any three states in the region $\rho > 0$, $-c < v < c$. Then*

$$|[u_L, u_R]| \leq |[u_L, u_M]| + |[u_M, u_R]|. \tag{4.12}$$

Proof: It was shown in [24] that (4.12) holds in the special relativistic case $divT_M = 0$. Since the effect of $\mathbf{A}_{ij}$ is to change the speeds of the elementary waves, but not the left and right states, in the solution of (4.1), (4.2), it follows that the estimate (4.12) continues to hold for arbitrary, (but constant), values of $\mathbf{A}_{ij}$. □

Proposition 4 is a direct consequence of the geometry of shock and rarefaction curves summarized above, and discussed further below, and is not true except in the special case $p = \sigma^2 \rho$, [24]. It follows from Proposition 4 that the only increase in the total variation of $\ln \rho_{\Delta x}(\cdot, t)$ in an approximate solution $u_{\Delta x}(\cdot, t)$ is due to increases that occur during the ODE steps (3.22). This is the basis for our analysis of convergence. Thus we analyze solutions in the z-plane, $\mathbf{z} = (z, w) \equiv (K_0 \ln \rho, \ln \frac{c-v}{c+v})$, a $45°$ rotation of the plane of Riemann invariants (r, s), c.f.(2.12), (2.13), [24].

Thus, let $\mathbf{z}_L, \mathbf{z}_R$ be the left and right states of a single elementary wave γ, and let γ denote both the *name* of the wave, as well as the *vector*

$$\gamma = \mathbf{z}_R - \mathbf{z}_L. \tag{4.13}$$

Let

$$\|\gamma\| = \|\mathbf{z}_R - \mathbf{z}_L\|, \tag{4.14}$$

and so we have

$$|\gamma| = |K_0 \ln \rho_R - K_0 \ln \rho_L| = |z_R - z_L| \leq \|\gamma\|, \tag{4.15}$$

where K_0 is defined in (2.14). Note that because changes in $\mathbf{A}$ affect only the speeds of waves, it follows that γ, $|\gamma|$ and $\|\gamma\|$ depend only on $\mathbf{z}_L, \mathbf{z}_R$, and not on the value of $\mathbf{A_{ij}}$ used in the construction. We write

$$[\mathbf{z}_L, \mathbf{z}_R] \equiv [u_L, u_R] = (\gamma^1, \gamma^2), \tag{4.16}$$

to indicate that γ^1, γ^2 are the elementary 1- and 2-waves that solve the Riemann Problem with left state $\mathbf{z}_L = \Phi \circ \Psi^{-1} u_L$ and right state $\mathbf{z}_L =$

$\Phi \circ \Psi^{-1} u_R$. We now review the results in [24] regarding the geometry of shock and rarefaction curves as plotted in the z-plane.

Let $S_i(\mathbf{z}_L)$ denote the i-shock curve emanating from the left state $\mathbf{z}_L$. That is, $\mathbf{z}_R \in S_i(\mathbf{z}_L)$ if and only if $[\mathbf{z}_L, \mathbf{z}_R]$ is a pure i-shock,[23]. It was shown in [24] that all i-shock curves are translates of one another in the z-plane, and 2-shock curves are just the reflection of the 1-shock curves about lines $z = const$. The following formula for the 1-shock curve is given in [24], p.85, equations (74),(75):

Lemma 1 *A state $\mathbf{z}_R$ lies on the 1-shock curve $S_1(\mathbf{z}_L)$ if and only if*

$$\Delta r = -\frac{1}{2}\ln\{f_+(2K\zeta)\} - \sqrt{\frac{K}{2}}\ln\{f_+(\zeta)\}, \qquad (4.17)$$

$$\Delta s = -\frac{1}{2}\ln\{f_+(2K\zeta)\} + \sqrt{\frac{K}{2}}\ln\{f_+(\zeta)\}, \qquad (4.18)$$

where

$$f_+(\zeta) = (1+\zeta) + \sqrt{\zeta(2+\zeta)}, \qquad (4.19)$$

for some $0 \leq \zeta < \infty$. Here

$$K = \frac{2\sigma^2 c^2}{(c^2+\sigma^2)^2} = \frac{K_0^2}{2}, \qquad (4.20)$$

is used in place of K_0, and $\Delta r = r_R - r_L$, $\Delta s = s_R - s_L$, denote the change in the Riemann invariants across the shock, c.f. [24].

Using (2.15),(2.16) we see that (4.17),(4.18) are equivalent to

$$\Delta w = -\ln\{f_+(K_0^2\zeta)\}, \qquad (4.21)$$
$$\Delta z = -K_0 \ln\{f_+(\zeta)\}. \qquad (4.22)$$

Since (4.21),(4.22) describe the 1-shock curves for $0 \leq \zeta < \infty$, it follows directly from these that 1-shock curves $S_1(\mathbf{z}_L)$ have a geometric shape in the z-plane that is independent of $\mathbf{z}_L$. Thus all 1-shock curves are translates of one another in the z-plane, as claimed.

It also was shown in [24] that the 2-shock curve $S_2(\mathbf{z}_L)$ is the reflection of $S_1(\mathbf{z}_L)$ about the line $z = z_L$, (this follows directly from (76), (77) of [24].) From this, together with (4.21),(4.22), it follows that

$$|\Delta w| = \ln\left\{f_+(K_0^2 \zeta)\right\}, \qquad (4.23)$$
$$|\Delta z| = K_0 \ln\left\{f_+(\zeta)\right\}, \qquad (4.24)$$

hold all along both the 1- and 2-shock curves. The next lemma implies the convexity of i-shock curves in the case $p = \sigma^2 \rho$.

Lemma 2 *The shock equations (4.23), (4.24) imply that*

$$\sinh\left(\frac{|\Delta w|}{2}\right) = K_0 \sinh\left(\frac{|\Delta z|}{2K_0}\right), \qquad (4.25)$$

from which it follows that (4.23), (4.24) define

$$|\Delta w| = H(|\Delta z|), \qquad (4.26)$$

where the function H is given by

$$H(|\Delta z|) = \ln f_+\left(2K_0^2 \sinh^2\left\{\frac{|\Delta z|}{2K_0}\right\}\right) = 2\sinh^{-1}\left(K_0 \sinh\frac{|\Delta z|}{2K_0}\right). \qquad (4.27)$$

The function H satisfies

$$H''(|\Delta z|) = \frac{(c^2 - \sigma^2)^2}{2c\sigma(c^2 + \sigma^2)} \frac{\sinh(\frac{|\Delta z|}{2K_0})}{\cosh^3(\frac{|\Delta w|}{2})} \geq 0. \qquad (4.28)$$

Proof: Solving equation (4.24) for ζ gives

$$\zeta = 2\left(\sinh\left(\frac{|\Delta z|}{2K_0}\right)\right)^2. \qquad (4.29)$$

Substituting (4.29) into (4.23) yields the first equality in (4.27), and the formula

$$f_+^{-1}(y) = 2\sinh^2(\ln y). \qquad (4.30)$$

Using this to solve for ζ in (4.23), (4.24), equating, and taking square roots, gives (4.25), as well as the second equality in (4.27). Implicitly differentiating (4.25) and simplifying gives (4.28). □

SHOCK-WAVE SOLUTIONS OF THE EINSTEIN EQUATIONS

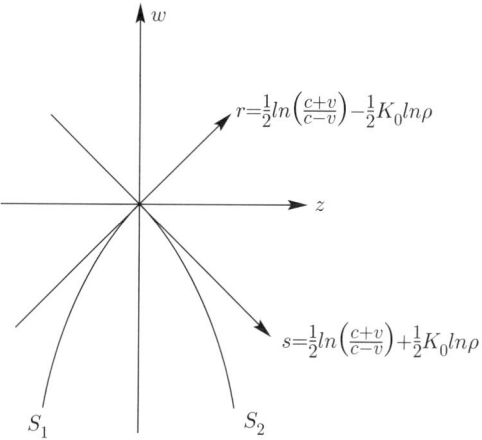

Figure 2: The reflection property of shock curves

It follows directly from Lemma 2 that $H(|\Delta z|)$ is a monotone increasing convex up function of $|\Delta z|$ that is superlinear in the sense that

$$|\Delta z| < H(|\Delta z|) < \infty, \tag{4.31}$$

for all $\Delta z \neq 0$, and

$$\lim_{|\Delta z| \to 0} \frac{H(|\Delta z|)}{|\Delta z|} = 1, \tag{4.32}$$

c.f. Figure 2,3.

Since $|\Delta w| = |\Delta z|$ along all 1- and 2-rarefaction curves, we have the following lemma:

Lemma 3 *Let $\mathbf{z}_L, \mathbf{z}_R$ be the left and right states of an elementary wave γ, so that*

$$\gamma = [\mathbf{z}_L, \mathbf{z}_R]. \tag{4.33}$$

Then

$$|\Delta w| \leq H(|\gamma|), \tag{4.34}$$

where

$$\Delta w = w_R - w_L, \tag{4.35}$$
$$|\gamma| = |\Delta z| = |z_R - z_L|, \tag{4.36}$$

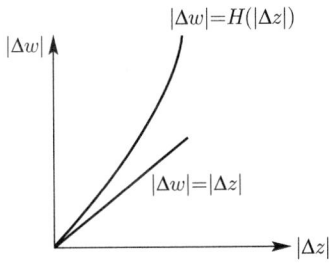

Figure 3: H is increasing convex up

where H is given in (4.26).

The superlinearity and convexity of H, together with Jenson's inequality, imply the following propostion:

Proposition 5 *Let $\gamma_1, ..., \gamma_n$ be any set of elementary waves. Then*

$$\sum_{i=1}^{n} |\gamma_i| \leq \sum_{i=1}^{n} H(|\gamma_i|) \leq H\left(\sum_{i=1}^{n} |\gamma_i|\right). \tag{4.37}$$

The next Proposition summarizes results in [24], and follows directly from Proposition 5:

Proposition 6 *For any left and right states $\mathbf{z}_L, \mathbf{z}_R \in R^2$, there exists a unique solution of the Riemann Problem $[\mathbf{z}_L, \mathbf{z}_R]$ consisting of a 1-shock or 1-rarefaction wave γ^1, followed by a 2-shock or 2-rarefaction wave γ^2, so that we can write*

$$(\gamma^1, \gamma^2) = [\mathbf{z}_L, \mathbf{z}_R]. \tag{4.38}$$

The speed of the wave γ^1 is always strictly less than the speed of γ^2, and all wave speeds are subluminous. Moreover, there exist C^2 functions $\Gamma_p : R^2 \to R^2$, one for each $p = 1, 2$, such that $(\gamma^1, \gamma^2) = [\mathbf{z}_L, \mathbf{z}_R]$ if and only if the vector γ^p satisfies

$$\gamma^p = \Gamma_p(\mathbf{z}_R - \mathbf{z}_L), \tag{4.39}$$

where,

$$|\gamma^p| = |\Gamma_p(\mathbf{z}_R - \mathbf{z}_L)| \leq \sqrt{2}\|\mathbf{z}_R - \mathbf{z}_L\|. \qquad (4.40)$$

Proof: The smoothness of Γ_p and the dependence on the difference $\mathbf{z}_R - \mathbf{z}_L$ follows from the C^2 contact between shock and rarefaction curves, together with the fact that shock-wave curves, drawn in the $\mathbf{z}$-plane, are translation invariant. Estimate (4.40) can be verified in each of the four cases of the Riemann Problem $[\mathbf{z}_L, \mathbf{z}_R]$; namely, if $[\mathbf{z}_L, \mathbf{z}_R]$ is a 1-shock followed by a 2-rarefaction wave *or* a 1-rarefaction wave followed by a 2-shock, then $|\gamma^1| + |\gamma^2| = |z_R - z_L|$. In the other two cases one can verify (4.40) assuming that the shock-waves lie on the Riemann Invariants, and then see that the divergences of shock and rarefaction curves only improves this estimate. $\square$

We now discuss the boundary Riemann problems posed at mesh points (x_0, t_j), $j = 0, 1, 2, ..$, that lie along the boundary $x_0 = r_0$ in the approximate solution $u_{\Delta x}$. In this case, for fixed j, $u_{0j}^{RP}(x, t)$ is defined in (3.8), (3.9) as the solution of (4.1) together with the initial-boundary data

$$u_0(x) = \left\{\begin{array}{ll} v = 0 & x = 0 \\ u_R = u_{0j} & x \geq 0 \end{array}\right\}, \qquad (4.41)$$

with the origin translated to the bottom center (x_0, t_j) of the mesh rectangle $\mathcal{R}_{0j} \equiv \{(x, t) : x_{r_0} < x \leq x_{\frac{1}{2}}, t_j \leq t < t_{j+1}\}$. Again, vector $\mathbf{A}_{0j}$ is constant on $\mathcal{R}_{0j}$. The following theorem, which generalizes Proposition 3 to include boundary Riemann problems, follows by similar reasoning. (See [21] for a discussion of boundary Riemann problems.)

Proposition 7 *Assume that u_R lies in the region $\rho > 0$, $-c < v < c$. Then the boundary Riemann problem (4.1), (4.41) has a unique solution consisting of a single elementary 2-wave γ_{0j}^2 of positive speed. Moreover, the CFL condition (3.2) guarantees that the speed of the wave γ_{0j}^2 is always smaller than half the the mesh speed $\frac{\Delta x}{\Delta t} = Max\left\{2\sqrt{\frac{A}{B}}\right\}$, and thus γ_{0j}^2 cannot hit the boundary of $\mathcal{R}_{0j}$ within one timestep.*

For fixed $\mathbf{A}_{0j}$, let

$$[0, u_R] \equiv [0, u_R](x, t), \qquad (4.42)$$

denote the solution of the Riemann problem (4.1), (4.41), and write

$$[0, u_R] = \gamma^2, \qquad (4.43)$$

to indicate that the solution $[0, u_R](x, t)$ consists of the single wave γ^2, a 2-wave. Again, define the strength of the Riemann problem $[0, u_R]$ as the strength of its elementary wave,

$$|[0, u_R]| = |\gamma^2|. \tag{4.44}$$

The following theorem generalizes Proposition 4 to include the boundary Riemann problems, and this implies that the sum of the strengths of elementary waves are non-increasing during boundary wave interactions, so long as $\mathbf{A}_{ij}$ is constant.

Proposition 8 *Assume that $\mathbf{A}_{ij}$ is fixed. Let u_M, and u_R be any pair of states in the region $\rho > 0$, $-c < v < c$. Then*

$$|[0, u_R]| \leq |[0, u_M]| + |[u_M, u_R]|. \tag{4.45}$$

5 The ODE Step

In this section we analyze the ODE step (3.22) of the fractional step scheme. Recall that this arises by rewriting system (1.30) in the form $u_t + \frac{\partial f}{\partial u} u_x = g - \mathbf{A}' \cdot \nabla_{\mathbf{A}} f \equiv G(\mathbf{A}, u, x)$ and neglecting the flux term containing u_x. Then the jumps in $\mathbf{A}$ at the vertical lines $x_{i+\frac{1}{2}}$, $i = 0, 1, ...$, are accounted for by the $\mathbf{A}' \cdot \nabla_{\mathbf{A}} f$ term on the RHS of this equation. Using (3.20), (3.21) and the fact that $(u^0, u^1) = (T_M^{00}, T_M^{01})$, system (3.22) takes the form

$$\dot{T}_M^{00} = -\frac{1}{2}\sqrt{\frac{A}{B}} T_M^{01} \left\{ \frac{2(B+1)}{x} - \kappa x B \left(T_M^{00} - T_M^{11} \right) \right\} \equiv G^0(\mathbf{A}, u, x), \tag{5.1}$$

$$\dot{T}_M^{00} = -\frac{1}{2}\sqrt{\frac{A}{B}} \left\{ \frac{4}{x} T_M^{11} + \frac{(B-1)}{x} \left(T_M^{00} + T_M^{11} \right) \right. \tag{5.2}$$

$$\left. +\kappa x B \left[T_M^{00} T_M^{11} - 2 \left(T_M^{01} \right)^2 + \left(T_M^{11} \right)^2 \right] - 4x T^{22} \right\} \equiv G^1(\mathbf{A}, u, x).$$

We now analyze the solution trajectories for system (5.1), (5.2) in the (ρ, v)-plane. To this end, we record the following identities which are easily derived from (1.15),(1.16),(1.17), and (2.6):

$$\left(T_M^{11} \right)^2 - \left(T_M^{01} \right)^2 = \frac{\sigma^4 - v^2 c^2}{c^2 - v^2} \rho^2 c^2, \tag{5.3}$$

$$T_M^{00}T_M^{11} - \left(T_M^{01}\right)^2 = \sigma^2\rho^2c^2, \tag{5.4}$$

$$\left(T_M^{11}\right)^2 - 2\left(T_M^{01}\right)^2 + T_M^{00}T_M^{11} = \frac{(\sigma^2 - v^2)(c^2 + \sigma^2)}{c^2 - v^2}\rho^2c^2, \tag{5.5}$$

$$T_M^{00} + T_M^{11} = \frac{(c^2 + \sigma^2)(c^2 + v^2)}{c^2 - v^2}\rho, \tag{5.6}$$

$$T_M^{00} - T_M^{11} = (c^2 - \sigma^2)\rho. \tag{5.7}$$

Using (5.3)-(5.7) in the RHS of (5.1), (5.2), we obtain after simplification,

$$G^0 = -\tfrac{1}{2}\sqrt{\tfrac{A}{B}}\left(\tfrac{c^2+\sigma^2}{c^2-v^2}\right)cv\tfrac{\rho}{x}\left\{2(B+1) - \kappa B(c^2 - \sigma^2)\rho x^2\right\}, \tag{5.8}$$

$$G^1 = -\tfrac{1}{2}\sqrt{\tfrac{A}{B}}\left(\tfrac{c^2+\sigma^2}{c^2-v^2}\right)\tfrac{\rho}{x}\left\{4v^2 + (B-1)(c^2+v^2) + \kappa B(\sigma^2 - v^2)c^2\rho x^2\right\}. \tag{5.9}$$

Now differentiating the LHS of (5.1), (5.2) gives

$$\dot{\rho}\frac{\partial T_M^{00}}{\partial \rho} + \dot{v}\frac{\partial T_M^{00}}{\partial v} = G^0, \tag{5.10}$$

$$\dot{\rho}\frac{\partial T_M^{01}}{\partial \rho} + \dot{v}\frac{\partial T_M^{01}}{\partial v} = G^1. \tag{5.11}$$

Thus it follows from Cramer's Rule that system (5.1), (5.2) in (ρ, v)-variables is given by

$$\dot{\rho} = \frac{D_\rho}{D}, \tag{5.12}$$

$$\dot{v} = \frac{D_v}{D}, \tag{5.13}$$

where

$$D_\rho = \begin{vmatrix} G^0 & \frac{\partial T_M^{00}}{\partial v} \\ G^1 & \frac{\partial T_M^{01}}{\partial v} \end{vmatrix}, \tag{5.14}$$

$$D_v = \begin{vmatrix} \frac{\partial T_M^{00}}{\partial \rho} & G^0 \\ \frac{\partial T_M^{01}}{\partial \rho} & G^1 \end{vmatrix}, \tag{5.15}$$

$$D = \begin{vmatrix} \frac{\partial T_M^{00}}{\partial \rho} & \frac{\partial T_M^{00}}{\partial v} \\ \frac{\partial T_M^{01}}{\partial \rho} & \frac{\partial T_M^{01}}{\partial v} \end{vmatrix}. \tag{5.16}$$

Using (1.15) and (1.16) we obtain

$$\frac{\partial T_M^{00}}{\partial \rho} = \frac{c^4 + \sigma^2 v^2}{c^2 - v^2},$$

$$\frac{\partial T_M^{00}}{\partial v} = 2\frac{(c^2 + \sigma^2)c^2 v}{(c^2 - v^2)^2}\rho,$$

$$\frac{\partial T_M^{01}}{\partial \rho} = \frac{(\sigma^2 + c^2)cv}{c^2 - v^2},$$

$$\frac{\partial T_M^{01}}{\partial v} = \frac{(\sigma^2 + c^2)(c^2 + v^2)c}{(c^2 - v^2)^2}\rho,$$

and

$$D = \frac{(c^2 + \sigma^2)(c^4 - \sigma^2 v^2)c}{(c^2 - v^2)^2}\rho. \tag{5.17}$$

A calculation using these together with (5.8), (5.9) leads to

$$D_\rho = -\frac{1}{2}\sqrt{\frac{A}{B}}\left(\frac{c^2 + \sigma^2}{c^2 - v^2}\right)^2 \frac{c^2}{x}\left\{4 - \kappa B(c^2 + \sigma^2)\rho x^2\right\}\rho^2,$$

$$D_v = -\frac{1}{2}\sqrt{\frac{A}{B}}\left(\frac{c^2 + \sigma^2}{c^2 - v^2}\right)\frac{\sigma^2 c^2}{x}$$
$$\times \left\{-4\frac{v^2}{c^2} + (B-1)\frac{c^4 - \sigma^2 v^2}{\sigma^2 c^2} + \kappa B(c^2 + v^2)\rho x^2\right\}\rho.$$

Putting (5.17) and the above expressions for D_ρ and D_v into (5.12), (5.13) and simplfying, we obtain system (5.1), (5.2) in (ρ, v)-variables:

$$\dot{\rho} = -\frac{1}{2}\sqrt{\frac{A}{B}}\left(\frac{c^2 + \sigma^2}{c^4 - \sigma^2 v^2}\right)\frac{vc}{x} \tag{5.18}$$
$$\times \left\{4 - \kappa B(c^2 + \sigma^2)\rho x^2\right\}\rho,$$

$$\dot{v} = -\frac{1}{2}\sqrt{\frac{A}{B}}\left(\frac{c^2 - v^2}{c^4 - \sigma^2 v^2}\right)\frac{\sigma^2 c}{x} \tag{5.19}$$
$$\times \left\{-4\frac{v^2}{c^2} + (B-1)\frac{c^4 - \sigma^2 v^2}{\sigma^2 c^2} + \kappa B(c^2 + v^2)\rho x^2\right\},$$

For convenience, we rewrite system (5.18), (5.19) in the form

$$\dot{\rho} = \frac{\kappa\sqrt{AB}x}{2}\left[\frac{(c^2+\sigma^2)^2 vc}{c^4-\sigma^2 v^2}\right]\rho\{\rho-\rho_1\}, \tag{5.20}$$

$$\dot{v} = -\frac{\kappa\sqrt{AB}x}{2}\left[\frac{(c^4-v^4)\sigma^2 c}{c^4-\sigma^2 v^2}\right]\{\rho-\rho_2\}, \tag{5.21}$$

where

$$\rho_1 = \frac{4}{\kappa B(c^2+\sigma^2)x^2}, \tag{5.22}$$

and

$$\rho_2 = \frac{4v^2\sigma^2 - (B-1)(c^4-\sigma^2 v^2)}{\kappa B(c^2+v^2)\sigma^2 c^2 x^2}, \tag{5.23}$$

where, (by a simple calculation),

$$\rho_2 < \frac{4v^2\sigma^2}{\kappa B(c^2+v^2)\sigma^2 c^2 x^2} < \rho_1, \tag{5.24}$$

for all values of $v \in (-c, c)$.

We devote the remainder of this section to the proof of the following theorem, which gives a global bound for solutions of $\dot{u} = G(\mathbf{A}, u, x)$, starting from arbitrary initial data

$$u(0) = u_0 \equiv \Psi(\rho_0, v_0), \tag{5.25}$$

assuming that $A > 0$, $B \geq 1$ and $x \geq r_0$ are constant, and assuming the physical bounds $0 < \rho_0 < \infty$, $-c < v_0 < c$, (c.f. (2.21)):

Proposition 9 *Assume that A, B and x are constant, that $A > 0$, $B \geq 1$, $x \geq r_0$, and assume that (ρ_0, v_0) satisfies $-c < v_0 < c$ and $0 < \rho_0 < \infty$. Then the solution $(\rho(t), v(t))$ of system (5.20), (5.21), with initial condition*

$$\rho(0) = \rho_0, \tag{5.26}$$
$$v(0) = v_0, \tag{5.27}$$

exists, is finite, and satisfies

$$-c < v(t) < c,$$

for all $t \geq 0$. Moreover, if $\rho_0 \leq \rho_1$, then $\rho(t) \leq \rho_1$ for all $t \geq 0$; while if $\rho_0 \geq \rho_1$, then

$$\rho_1 \leq \rho(t) \leq \max\left\{\rho_1, \rho_0 \left(\frac{c^2}{c^2 - v_0^2}\right)^{\frac{1}{2}\left(\frac{c^2+\sigma^2}{c\sigma}\right)^2}\right\}, \tag{5.28}$$

for all $t > 0$.

Proof: For fixed $\mathbf{A}$ and x, system (5.20), (5.21) is an autonomous system of the form

$$\dot{\rho} = f_1(\rho, v),$$
$$\dot{v} = f_2(\rho, v).$$

Note that $\rho = 0$ and $v = \pm c$ are solution trajectories for this system. Since the system is autonomous, solution trajectories never intersect, and so it follows that $\rho > 0$, $|v| < c$ is an invariant region for solutions. Note also that since ρ_1 is independent of v, the isocline $\rho \equiv \rho_1$ also defines a solution curve for system (5.20), (5.21), and so it also cannot be crossed by other solution trajectories. Thus $0 < \rho < \rho_1$, $|v| < c$ is a bounded invariant region, and $\rho > \rho_1$, $|v| < c$ is an unbounded invariant region, for solutions of system (5.20), (5.21). Thus it remains only to verify (5.28), and it follows that the only obstacle to global existence for the initial value problem (5.20), (5.21), (5.26), (5.27), is the case $\rho_0 > \rho_1$, and the possibility that $\rho(t) \to \infty$ before $t \to \infty$. Note that (5.20) is quadratic in ρ, so the bound (5.28) on ρ is not a consequence of equation (5.20) alone. However, (5.21) implies that ρ is bounded, as we now show.

If $\rho_0 > \rho_1$, then since $\rho_1 > \rho_2$ for all values of v, it follows that $\dot{v} < 0$ for all time. Consequently, $v(t) \leq v_0$, and $\rho(t)$ can only increase while $v \geq 0$. Once v hits $v = 0$, $v(t) < 0$ and $\rho(t)$ decreases from that time forward. Thus it suffices to estimate the change in $\rho(t)$ while $0 \leq v(t) \leq v_0$. But from (5.20), (5.21), we have

$$\frac{d\rho}{dv} = -\frac{(c^2 + \sigma^2)^2 v}{(c^4 - v^4)\sigma^2} \frac{\rho - \rho_1}{\rho - \rho_2}\rho \tag{5.29}$$

$$\geq -\left(\frac{c^2 - \sigma^2}{\sigma c}\right)^2 \frac{v}{c^2 - v^2}\rho,$$

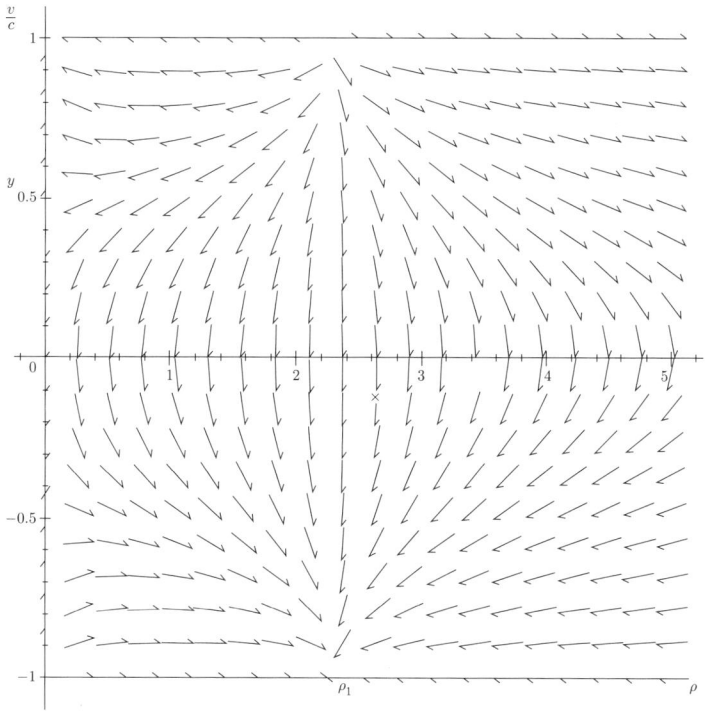

Figure 4: The phase portrait for system (5.20), (5.21)

where we have used $\rho \geq \rho_1 \geq \rho_2$. Integrating this inequality by separation of variables gives the inequality (5.28). □

The phase portrait for solutions of (5.20), (5.21), is given in Figure 4.

By the results of Section 4 the Riemann problem solutions preserve the bounds $0 < \rho < \infty$, $|v| < c$, and all (invariant) wave speeds remain bounded by c, so long as $0 < \rho < \infty$, $|v| < c$ initially. By the results in this section, it follows that these bounds also are maintained under the ODE step. But (3.26) and (3.43) imply that that the only way the approximate solution $u_{\Delta x}$ can fail to be defined for all time, is if $B \to \infty$, or the CFL condition fails. The following theorem is a direct consequence of (3.26) and (3.43):

Proposition 10 *Let $\bar{B}$, $\bar{M}$ denote arbitrary positive constants, let*

$$\Lambda = \frac{\Delta x}{\Delta t} = 2\sqrt{G_{AB}(\bar{B}, \bar{M})},$$

and assume that the initial data $u_0(\cdot)$ satisfies the bounds $0 < \rho < \infty$, $|v| < c$ for all $x \geq r_0$. Then the approximate solution $u_{\Delta x}$ is defined, and continues to satisfy the bounds $0 < \rho < \infty$, $|v| < c$ for all $x \geq r_0$, $t \leq t_J$, so long as

$$\|M_{\Delta x}\|_\infty < \bar{M},$$

$$\|B_{\Delta x}\|_\infty < \bar{B},$$

for all $x \geq r_0$, $j \leq J$.

As a final comment, note that we have bounds for the RP step, and bounds for the ODE step, but it remains to obtain bounds that apply to both steps. Also, the fact that ρ and v remain finite in each approximate solution does not rule out $\rho \to \infty$ in the actual solution. For this, we need estimates that are independent of Δx, c.f. [16].

6   Estimates for the ODE step

In this section we obtain estimates for the growth of the total variation of $\ln \rho$ and $\ln \frac{c+v}{c-v}$ under the evolution of the ODE $\dot{u} = G(\mathbf{A}, r, x)$, which is equivalent to the system (5.20), (5.21). To this end, rewrite system (5.20), (5.21) in terms of the variables $(z, w) \equiv (K_0 \ln \rho, K_0 \ln \frac{c+v}{c-v})$ to obtain, c.f. (2.15), (2.16),

$$\begin{aligned}
\dot{z} &= \frac{4\sqrt{AB}}{x}\left(\frac{\sigma v c^2}{c^4 - \sigma^2 v^2}\right)\left\{\frac{\kappa(c^2 + \sigma^2)}{4}\rho x^2 - \left(\frac{1}{B}\right)_1\right\} \quad (6.1)\\
&\equiv F_1(A, B, x, z, w)\\
\dot{w} &= \frac{4\sqrt{AB}}{x}\left(\frac{c^4}{c^4 - \sigma^2 v^2}\right)\left\{\frac{\kappa(c^2 + v^2)}{4}\frac{\sigma^2}{c^2}\rho x^2 \right.\quad (6.2)\\
&\qquad \left. - \left(\frac{\sigma^2}{c^2}\left[\frac{1}{B}\frac{v^2}{c^2} - \frac{(B-1)}{4B}\frac{c^4 - \sigma^2 v^2}{\sigma^2 c^2}\right]\right)_2\right\}\\
&\equiv F_2(A, B, x, z, w).
\end{aligned}$$

Here K_0 is defined in (2.14), and we use that

$$\dot{z} = K_0 \frac{\dot{\rho}}{\rho},$$
$$\dot{w} = \frac{2c}{c^2 - v^2}\dot{v}.$$

Note also that $\kappa c^2 \rho x^2$ is dimensionless. A calculation shows that the indexed brackets on the RHS of (6.1), (6.2) satisfy

$$|(\cdot)_i| \leq 1, \quad i = 1, 2. \tag{6.3}$$

To verify (6.3), use that $B \geq 1$, and

$$\begin{aligned}
|(\cdot)_2| &= \frac{\sigma^2}{c^2} \left| \frac{(c^4 + 3\sigma^2 v^2) - B(c^4 - \sigma^2 v^2)}{4B\sigma^2 c^2} \right| \\
&= \frac{\sigma^2}{c^2} \left| \frac{1}{4}\left(\frac{1}{B} - 1\right)\left(\frac{c}{\sigma}\right)^2 + \frac{1}{4}\left(\frac{3}{B} + 1\right)\left(\frac{v}{c}\right)^2 \right| \\
&\leq \frac{\sigma^2}{c^2} \text{Max}\left\{ \frac{1}{4}\left(\frac{c}{\sigma}\right)^2, \left(\frac{v}{c}\right)^2 \right\} \leq 1.
\end{aligned}$$

The following theorem gives bounds for the RHS of (6.1), (6.2).

Proposition 11 *Assume that*

$$1 \leq B \leq \bar{B}, \tag{6.4}$$
$$0 < AB \leq G_{AB}(\bar{B}, \bar{M}) \equiv A_{r_0} B_{r_0} \exp\left\{\frac{8\bar{B}\bar{M}}{r_0}\right\}, \tag{6.5}$$
$$S \leq \bar{S}, \tag{6.6}$$
$$|v| < c, \tag{6.7}$$

and $r_0 \leq x < \infty$. Then each of

$$|F_i(A, B, x, z, w)|, \quad \left|\frac{\partial F_i}{\partial z}\right|, \quad \left|\frac{\partial F_i}{\partial w}\right|,$$

$i = 1, 2$, *is bounded by* $\frac{1}{2\sqrt{2}} G_1(\bar{B}, \bar{M}, \bar{S})$, *where G_1 is defined by*

$$\frac{1}{2\sqrt{2}} G_1(\bar{B}, \bar{M}, \bar{S}) \equiv \frac{G_0(\kappa c^2 r_0 \bar{S} + 1)}{r_0}, \tag{6.8}$$

where

$$G_0 \equiv G_0(\bar{B}, \bar{M}) = K_1 \sqrt{G_{AB}(\bar{B}, \bar{M})}, \tag{6.9}$$

$$K_1 = \frac{8c^4}{(c^2 - \sigma^2)^2}. \tag{6.10}$$

Here we use the notation that K with a subscript denotes a constant that depends only on κ, r_0, σ and c, while $G(\cdot)$ denotes a constant that depends also on $\bar{A}, \bar{B}, \bar{S}$ and $\bar{M}$, whichever appear in the parentheses after the G. We include the factor $(2\sqrt{2})^{-1}$ in (6.8) for future convenience, c.f. Theorem 12 and Proposition 13 below.

Proof: This follows by direct calculation, using $|v| < c$, $\sigma < c$. For example,

$$\begin{aligned}\left|\frac{\partial F_2}{\partial z}\right| &= \left|\frac{\partial \rho}{\partial z}\frac{\partial F_2}{\partial \rho}\right| = \left|\frac{\rho}{K_0}\frac{\partial F_2}{\partial \rho}\right| \\ &= \left|\frac{\rho(c^2 + v^2)}{2\sigma c}\frac{4\sqrt{AB}}{x}\left(\frac{c^4}{c^4 - \sigma^2 v^2}\right)\left\{\frac{\kappa(c^2 + v^2)}{4}\frac{\sigma^2}{c^2}x^2\right\}\right| \\ &\leq \frac{G_0(\bar{B}, \bar{M})}{r_0}(\kappa c^2 r_0 \bar{S} + 1).\end{aligned}$$

Also,

$$\begin{aligned}\left|\frac{\partial F_2}{\partial w}\right| &= \left|\frac{\partial v}{\partial w}\frac{\partial F_2}{\partial v}\right| = \left|\left(\frac{c^2 - v^2}{2c}\right)\frac{\partial F_2}{\partial v}\right| \\ &= \left|\left\{\frac{4\sqrt{AB}}{x}\left(\frac{c^2 - v^2}{2c}\right)\frac{\partial}{\partial v}\left(\frac{c^4}{c^4 - \sigma^2 v^2}\right)\{\cdot\}_*\right\}_3 \right. \\ &\quad \left. + \left\{\frac{4\sqrt{AB}}{x}\left(\frac{c^3(c^2 - v^2)}{2(c^4 - \sigma^2 v^2)}\right)\frac{\partial}{\partial v}\{\cdot\}_*\right\}_4\right|,\end{aligned}$$

where

$$\{\cdot\}_* = \left\{\frac{\kappa(c^2 + v^2)}{4}\frac{\sigma^2}{c^2}\rho x^2 - \left(\frac{\sigma^2}{c^2}\left[\frac{1}{B}\frac{v^2}{c^2} - \frac{(B-1)}{4B}\frac{c^4 - \sigma^2 v^2}{\sigma^2 c^2}\right]\right)_2\right\}_*.$$

But straightforward estimates show that

$$|\{\cdot\}_i| \leq \frac{1}{2}\frac{G_0(\bar{B}, \bar{M})}{r_0}(\kappa c^2 r_0 \bar{S} + 1),$$

for both $i = 1$ and $i = 2$, and so

$$\left|\frac{\partial F_2}{\partial w}\right| \leq \frac{G_0(\bar{B}, \bar{M})}{r_0}(\kappa c^2 r_0 \bar{S} + 1).$$

This completes the proof of the theorem.□

We now study solutions of (6.1), (6.2) in the **z**-plane,

$$\mathbf{z} = (z, w) \equiv (K_0 \ln \rho, \ln \frac{c-v}{c+v}), \tag{6.11}$$

$$\|\mathbf{z}\| = \sqrt{z^2 + w^2}, \tag{6.12}$$

so that system (6.1), (6.2) can be written as

$$\dot{\mathbf{z}} = F(\mathbf{A}, x, \mathbf{z}), \tag{6.13}$$

where $\mathbf{A} = (A, B)$ and

$$F = (F_1, F_2). \tag{6.14}$$

Let

$$\mathbf{z}(t) \equiv \mathbf{z}(t; \mathbf{A}, x, \mathbf{z}_0) \tag{6.15}$$

denote the solution of (6.1), (6.2) starting from initial data

$$\mathbf{z}(0) = \mathbf{z}_0, \tag{6.16}$$

treating $\mathbf{A}$ and x as constants. We now estimate

$$\frac{d}{dt}\|\mathbf{z}(t)\|. \tag{6.17}$$

To start, note first that for any smooth curve $\mathbf{z}(t)$,

$$\left|\frac{d}{dt}\|\mathbf{z}\|\right| = \left|\frac{\mathbf{z}(t) \cdot \dot{\mathbf{z}}(t)}{\|\mathbf{z}(t)\|}\right| \leq \|\dot{\mathbf{z}}(t)\|. \tag{6.18}$$

Thus, if $\mathbf{z}(t)$ denotes a solution of (6.1), (6.2), then

$$\begin{aligned}\left|\frac{d}{dt}\|\mathbf{z}\|\right| &\leq \|F(A, B, x, \mathbf{z}(t))\| \\ &= \sqrt{2}\frac{G_0(\bar{B}, \bar{M})}{r_0}(\kappa c^2 r_0 \bar{S} + 1).\end{aligned} \tag{6.19}$$

We next obtain a similar estimate for

$$\left|\frac{d}{dt}\|\mathbf{z}_R(t) - \mathbf{z}_L(t)\|\right|, \tag{6.20}$$

where

$$\mathbf{z}_L(t) \equiv \mathbf{z}(t; \mathbf{A}, x_L, \mathbf{z}_L), \tag{6.21}$$
$$\mathbf{z}_R(t) \equiv \mathbf{z}(t; \mathbf{A}, x_R, \mathbf{z}_R), \tag{6.22}$$

and $\mathbf{A}_L, x_L, \mathbf{A}_R, x_R$, are constants. (Here, $\mathbf{z}_L, \mathbf{z}_R$ could be consecutive constant states that pose a Riemann problem in the construction of $u_{\Delta x}$.) Then,

$$\begin{aligned}
\left|\frac{d}{dt}\|\mathbf{z}_R(t) - \mathbf{z}_L(t)\|\right| &\leq \|\dot{\mathbf{z}}_R(t) - \dot{\mathbf{z}}_L(t)\| \\
&= \|F(\mathbf{z}_R, \mathbf{A}, x) - F(\mathbf{z}_L, \mathbf{A}, x)\| = \|\Delta F\| \\
&\leq \sqrt{2} Max\{|\Delta F_1|, |\Delta F_2|\}.
\end{aligned} \tag{6.23}$$

But

$$|\Delta F_i| \leq \left|\frac{\partial F_i}{\partial z}\right| |\Delta z| + \left|\frac{\partial F_i}{\partial w}\right| |\Delta w|, \tag{6.24}$$

so by Proposition 11, if (6.4)-(6.7) hold, then

$$\begin{aligned}
|\Delta F_i| &\leq \frac{G_0(\bar{B}, \bar{M})}{r_0}(\kappa c^2 r_0 \bar{S} + 1)\{|\Delta z| + |\Delta w|\} \\
&\leq \sqrt{2}\frac{G_0(\bar{B}, \bar{M})}{r_0}(\kappa c^2 r_0 \bar{S} + 1)\{\|\Delta \mathbf{z}\|\}.
\end{aligned} \tag{6.25}$$

We have the following result:

Proposition 12 *Let $\mathbf{z}_L(t), \mathbf{z}_R(t)$ be defined by (6.21) and (6.22), and assume $x_L < x_R$, and that (6.4)-(6.7) of Proposition 11 hold. Then*

$$\left|\frac{d}{dt}\|\mathbf{z}_R(t) - \mathbf{z}_L(t)\|\right| \leq \frac{G_1}{\sqrt{2}}\{\|\mathbf{z}_R(t) - \mathbf{z}_L(t)\|\}, \tag{6.26}$$

$$\|\mathbf{z}_R(t) - \mathbf{z}_L(t)\| \leq \|\mathbf{z}_R - \mathbf{z}_L\| e^{\frac{G_1}{\sqrt{2}}t}, \tag{6.27}$$

where, c.f. (6.8),

$$G_1 \equiv G_1(\bar{B}, \bar{M}, \bar{S}) = 2\sqrt{2}\frac{G_0(\bar{B}, \bar{M})}{r_0}(\kappa c^2 r_0 \bar{S} + 1). \tag{6.28}$$

The states $\mathbf{z}_L(t), \mathbf{z}_R(t)$ pose a Riemann Problem $[\mathbf{z}_L(t), \mathbf{z}_R(t)]$ at each time $t \geq 0$. Let

$$[\mathbf{z}_L(t), \mathbf{z}_R(t)] = (\gamma^1(t), \gamma^2(t)), \tag{6.29}$$

denote the waves that solve this Riemann problem, c.f. (4.16).

Lemma 4 *The following estimate holds:*

$$\left| \frac{d}{dt} \sum_{p=1,2} |\gamma^p(t)| \right| \leq \sqrt{2} \|\dot{\mathbf{z}}_R(t) - \dot{\mathbf{z}}_L(t)\|. \tag{6.30}$$
$$\leq G_1 \|\Delta \mathbf{z}\|.$$

Proof: For the purposes of the proof, let $\mathbf{z}(t) = \mathbf{z}_R(t) - \mathbf{z}_L(t)$, and let

$$\gamma^p(\mathbf{z}(t)) \equiv \Gamma_p(\mathbf{z}(t)), \tag{6.31}$$

where Γ_p is defined in (4.39) of Proposition 6. Then by Propositions 4, 8 and (4.40),

$$\sum_{p=1,2} \{|\gamma^p(\mathbf{z}(t))| - |\gamma^p(\mathbf{z}(0))|\} \leq \sum_{p=1,2} |\gamma^p(\mathbf{z}(t) - \mathbf{z}(0))|$$
$$\leq \sqrt{2}\|\mathbf{z}(t) - \mathbf{z}(0)\|. \tag{6.32}$$

Similarly,

$$\sum_{p=1,2} \{|\gamma^p(\mathbf{z}(0))| - |\gamma^p(\mathbf{z}(t))|\} \leq \sum_{p=1,2} |\gamma^p(\mathbf{z}(0) - \mathbf{z}(t))|$$
$$\leq \sqrt{2}\|\mathbf{z}(t) - \mathbf{z}(0)\|. \tag{6.33}$$

Thus (6.32) together with (6.33) imply that

$$\left| \frac{d}{dt} \sum_{p=1,2} |\gamma^p(\mathbf{z}(t))| \right| \leq \sqrt{2}\|\dot{\mathbf{z}}(t)\|, \tag{6.34}$$

which is (6.30). The second inequality in (6.30) follows directly from (6.26). $\square$

We have the following, c.f. (6.8):

Proposition 13 *Let $(\gamma^1(t), \gamma^2(t)) = [\mathbf{z}_L(t), \mathbf{z}_R(t)]$, where $\mathbf{z}_L(t), \mathbf{z}_R(t)$ solve the ODE (6.1), (6.2). Assume that (6.4)-(6.7) of Proposition 11 hold. Then*

$$|\gamma^1(t)| + |\gamma^2(t)| \leq |\gamma^1(0)| + |\gamma^2(0)| + \|\mathbf{z}_R(0) - \mathbf{z}_L(0)\| e^{G_1 t} G_1 t, \qquad (6.35)$$

where $G_1 \equiv G_1(\bar{B}, \bar{M}, \bar{S})$ is given in (6.28).

Proof: This follows from (6.30) and (6.27). □

7 Analysis of the Approximate Solutions

Let $u_{\Delta x}(x, t)$ denote an approximate solution generated by the fractional step Glimm method, starting from initial data $u_0(\cdot)$ that satisfies the finite total mass condition

$$M_{\Delta x}(\infty, 0) = M_{r_0} + M_0 \leq \infty, \quad M_0 \equiv \frac{\kappa}{2} \int_{r_0}^{\infty} u^0_{\Delta x}(r, 0) r^2 \, dr; \qquad (7.1)$$

the condition for initial locally finite total variation,

$$\sum_{\substack{i_1 \leq i \leq i_2 \\ p=1,2}} |\gamma^p_{i,0}| < V_0, \qquad (7.2)$$

for all i_1, i_2 such that $|x_{i_2} - x_{i_1}| \leq L$; the condition that the initial velocity is bounded uniformly away from the speed of light,

$$|v_{\Delta x}(x, 0)| \leq \bar{v}_0 < c; \qquad (7.3)$$

and the condition that the initial supnorm of $x\rho$ is bounded,

$$S_{\Delta x}(x, 0) \equiv x \rho_{\Delta x}(x, 0) \leq \bar{S}_0 < \infty. \qquad (7.4)$$

Note that (7.1) and (7.4) imply that

$$|w_{\Delta x}(x, 0)| \leq \left|\ln\left(\frac{c + \bar{v}_0}{c - \bar{v}_0}\right)\right| \equiv \bar{w}_0, \qquad (7.5)$$

and

$$|\bar{z}_{\Delta x}(x, 0)| \leq \left|K_0 \ln\left(\frac{\bar{S}}{r_0}\right)\right| \equiv \bar{z}_0. \qquad (7.6)$$

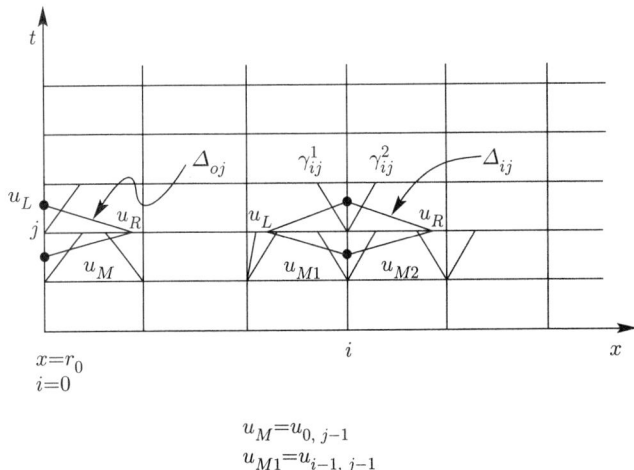

Figure 5: The interaction diamonds Δ_{ij}

$$\hat{u}_L = u_{i-1,j}^{RP},$$
$$\hat{u}_R = u_{ij}^{RP}. \quad (7.9)$$

States u_L, u_R are obtained from $\hat{u}_L, \hat{u}_R$ by solving the ODE $u_t = G$, written out in (3.19). That is, $u_L = \chi^{-1}(\mathbf{z}_L), u_R = \chi^{-1}(\mathbf{z}_R)$, and $\hat{u}_L = \chi^{-1}(\hat{\mathbf{z}}_L), \hat{u}_R = \chi^{-1}(\hat{\mathbf{z}}_R)$, where,

$$\mathbf{z}_L = \mathbf{z}(\Delta t; \mathbf{A}_{ij}, x_i, \hat{\mathbf{z}}_L), \quad (7.10)$$
$$\mathbf{z}_R = \mathbf{z}(\Delta t; \mathbf{A}_{ij}, x_i, \hat{\mathbf{z}}_R),$$

c.f. (6.15). Thus, using the notation introduced at (4.16),

$$[u_L, u_R] = (\gamma_{ij}^1, \gamma_{ij}^2),$$
$$[\hat{u}_L, \hat{u}_R] = (\hat{\gamma}_{ij}^1, \hat{\gamma}_{ij}^2), \quad (7.11)$$
$$[u_{M_1}, u_{M_2}] = (\gamma_{i,j-1}^1, \gamma_{i,j-1}^2),$$

and we write

$$[\hat{u}_L, u_{M_1}] = (\gamma_{i-1,j-1}^{R1}, \gamma_{i-1,j-1}^{R2}), \quad (7.12)$$
$$[u_{M_2}, u_R] = (\gamma_{i+1,j-1}^{L1}, \gamma_{i+1,j-1}^{L2}).$$

SHOCK-WAVE SOLUTIONS OF THE EINSTEIN EQUATIONS

Assuming (7.1)-(7.6), our goal is to find estimates for V_j, $\bar{S}_j$ and $T > 0$ such that

$$\sum_{\substack{i_1 \leq i \leq i_2 \\ p = 1,2}} |\gamma_{ij}^p| < V_j, \tag{7.7}$$

and

$$S_j \equiv \sup_{r \geq r_0} x \rho_{\Delta x}(x, t_j) \leq \bar{S}_j \tag{7.8}$$

for all $|x_{i_2} - x_{i_1}| \leq L$, $0 \leq t_j \leq T = t_J \leq 1$. Note that (7.7) estimates the total variation in z on x-intervals of length L, and (7.8) estimates a weighted supnorm. Estimates for the supnorm and local total variation norm of the approximate solution $u_{\Delta x}$, that are uniform in time, are required to apply the Oleinik compactness argument, [16]. Recall that the waves γ_{ij}^p solve the Riemann Problem $[u_{i-1,j}, u_{ij}]$ for system (3.8), and that since $\mathbf{A} = \mathbf{A_{ij}}$ on mesh rectangle $\mathcal{R}_{ij}$, it follows that the source $\mathbf{A}$ affects the speeds of the waves γ_{ij}^p, but the states on the waves themselves agree with the solution $[u_{i-1,j}, u_{ij}]$ for the special relativistic Euler equations (3.8) when $\mathbf{A} = (1,1)$.

To start, let Δ_{ij} denote the interaction diamond centered at (x_i, t_j) in the approximate solution $u_{\Delta x}$. In the case $i > 0$, the diamond Δ_{ij} is formed by the points $(x_{i-1} + a_j \Delta x, t_j)$, $(x_i + a_j \Delta x, t_j)$, $(x_i, t_{j-\frac{1}{2}})$, $(x_i, t_{j+\frac{1}{2}})$, and in the case $i = 0$, $\Delta_{0,j}$ is the half-diamond formed at the boundary by the mesh points $(x_0, t_{j+\frac{1}{2}})$, $(x_0, t_{j-\frac{1}{2}})$, $(x_0 + a_j \Delta x, t_j)$, c.f. Figure 5.

In the case $i > 0$, the waves γ_{ij}^p solve the Riemann Problem $[u_L, u_R]$, where

$$u_L = u_{i-1,j},$$
$$u_R = u_{ij}.$$

We call these the waves that leave the diamond Δ_{ij}, c.f. [8]. The waves that enter the diamond solve the Riemann Problems $[\hat{u}_L, u_{M_1}]$, $[u_{M_1}, u_{M_2}]$, and $[u_{M_2}, \hat{u}_R]$, where

$$u_{M_1} = u_{i-1,j-1},$$
$$u_{M_2} = u_{i,j-1},$$

and

Here we let $\gamma_{ij}^{R1}, \gamma_{ij}^{R2}$ denote the waves in the Riemann Problem posed at (x_i, t_j) that lie to the right of the random point $(x_i + a_{j+1}\Delta x, t_{j+1})$ at time $t = t_{j+1}$, and $\gamma_{ij}^{L1}, \gamma_{ij}^{L2}$ the waves that fall to the left of the random point $(x_{i-1} + a_{j+1}\Delta x, t_{j+1})$, respectively, [23, 8]

In the case of the boundary diamond $\Delta_{0,j}$, the wave γ_{0j}^2 leaves the diamond $\Delta_{0,j}$, and the waves $\gamma_{0,j-1}^2, \gamma_{1,j-1}^{L1}$ and $\gamma_{1,j-1}^{L2}$ enter the diamond. In this case, using the notation introduced at (4.44), (4.43), we can write

$$\begin{aligned} [0, u_R] &= \gamma_{0j}^2, \\ [0, \hat{u}_M] &= \hat{\gamma}_{0,j-1}^2, \\ [u_M, \hat{u}_R] &= (\gamma_{1,j-1}^{L1}, \gamma_{1,j-1}^{L2}), \end{aligned} \quad (7.13)$$

where

$$\begin{aligned} u_R &= u_{0j}, \\ \hat{u}_R &= u_{0j}^{RP}, \\ u_M &= u_{0,j-1}. \end{aligned} \quad (7.14)$$

Now let $|\gamma_{ij}^{IN}|$ denote the sum of the strength of the waves that enter the diamond Δ_{ij}. Thus, if $i > 0$, then

$$|\gamma_{ij}^{IN}| = \sum_{p=1,2} \left\{ |\gamma_{i-1,j-1}^p| + |\gamma_{i,j-1}^p| + |\gamma_{i+1,j-1}^{Lp}| \right\}, \quad (7.15)$$

and if $i = 0$,

$$|\gamma_{0j}^{IN}| = |\gamma_{0,j-1}^2| + |\gamma_{1,j}^{L1}| + |\gamma_{1,j}^{L2}|. \quad (7.16)$$

It follows from Propositions 4 and 8 that when $i > 0$,

$$|\hat{\gamma}_{ij}^1| + |\hat{\gamma}_{ij}^2| \leq |\gamma_{ij}^{IN}|, \quad (7.17)$$

and when $i = 0$,

$$|\hat{\gamma}_{0j}^2| \leq |\gamma_{0j}^{IN}|. \quad (7.18)$$

Now it follows from (6.35) of Proposition 13, that if $u_{\Delta x}$ satisfies (6.4)-(6.7) of Proposition 11, then

$$|\gamma_{ij}^1| + |\gamma_{ij}^2| \leq |\hat{\gamma}_{ij}^1| + |\hat{\gamma}_{ij}^2| + \|\hat{\mathbf{z}}_R - \hat{\mathbf{z}}_L\| e^{G_1 \Delta t} G_1 \Delta t. \quad (7.19)$$

But by (4.26),

$$|\hat{\mathbf{z}}_R - \hat{\mathbf{z}}_L| \leq |\hat{\gamma}_{ij}^1| + |\hat{\gamma}_{ij}^2| + H(|\hat{\gamma}_{ij}^1|) + H(|\hat{\gamma}_{ij}^2|), \tag{7.20}$$

so putting these together we obtain,

$$|\gamma_{ij}^1| + |\gamma_{ij}^2| \leq |\hat{\gamma}_{ij}^1| + |\hat{\gamma}_{ij}^2| + \left\{\sum_{p=1,2}\left[|\hat{\gamma}_{ij}^p| + H(|\hat{\gamma}_{ij}^p|)\right]\right\}e^{G_1 \Delta t}G_1 \Delta t. \tag{7.21}$$

We can now use (7.17), (7.18) to estimate $|\hat{\gamma}_{ij}^p|$ and $H(|\hat{\gamma}_{ij}^p|)$.

Note first that by the convexity of H, we have Proposition 5, so

$$\sum_{p=1,2} H\left(|\hat{\gamma}_{ij}^p|\right) \leq H\left(\sum_{p=1,2}|\hat{\gamma}_{ij}^p|\right) \leq H(|\gamma_{ij}^{IN}|). \tag{7.22}$$

Let

$$|\gamma_{ij}^{OUT}| = |\gamma_{ij}^1| + |\gamma_{ij}^2|. \tag{7.23}$$

Then putting (7.17), (or (7.18) at the boundary), and (7.22) into (7.21), we obtain

$$|\gamma_{ij}^{OUT}| \leq |\gamma_{ij}^{IN}| + \left\{|\gamma_{ij}^{IN}| + H\left(|\gamma_{ij}^{IN}|\right)\right\}e^{G_1 \Delta t}G_1 \Delta t. \tag{7.24}$$

We can also estimate the change in z and $\mathbf{z}$ between (x_i, t_j) and (x_i, t_{j-1}). Since both $\mathbf{z}_{i,j-1}$ and $\mathbf{z}_{ij}^{RP}$ are states on the waves $\gamma_{i,j-1}^p$ or $\gamma_{i+1,j-1}^p$, and by (6.19) we know

$$|z_{ij} - z_{ij}^{RP}| \leq \|\mathbf{z}_{ij} - \mathbf{z}_{ij}^{RP}\| \leq G_1 \Delta t, \tag{7.25}$$

it follows that

$$|\mathbf{z}_{ij} - \mathbf{z}_{i,j-1}| \leq \sum_{\substack{l=i,i+1 \\ p=1,2}} \left\{|\gamma_{l,j-1}^p| + H\left(|\gamma_{l,j-1}^p|\right)\right\} + G_1 \Delta t, \tag{7.26}$$

and

$$|z_{ij} - z_{i,j-1}| \leq \sum_{\substack{l=i,i+1 \\ p=1,2}} |\gamma_{l,j-1}| + G_1 \Delta t. \tag{7.27}$$

We collect our results so far in the following theorem.

SHOCK-WAVE SOLUTIONS OF THE EINSTEIN EQUATIONS

Theorem 4 *Let $\bar{M}$, $\bar{B}$, $\bar{S}$, $\bar{v}$, and integer $J_0 > 0$, be any finite positive constants, assume $|\bar{v}| < c$, and let $u_{\Delta x}(x,t)$, $\mathbf{A}_{\Delta x}(x,t)$ be an approximate solution generated by the fractional step Glimm method with*

$$\frac{\Delta x}{\Delta t} = \Lambda = 2\sqrt{G_{AB}(\bar{B},\bar{M})}. \tag{7.28}$$

Assume that,

$$M_{\Delta x}(x,t_j) \leq \bar{M}, \tag{7.29}$$
$$B_{\Delta x}(x,t_j) \leq \bar{B}, \tag{7.30}$$
$$0 < S_{\Delta x}(x,t_j) \leq \bar{S} \tag{7.31}$$
$$|v_{\Delta x}(x,t_j)| \leq \bar{v}, \tag{7.32}$$

for all $x \geq r_0$, $0 \leq t_j \leq T_0 = t_{J_0} \leq 1$. Then the speed of each wave γ_{ij}^p generated in $u_{\Delta x}$ at the Riemann Problem step of the method is bounded by the coordinate speed of light $\sqrt{A_{ij}/B_{ij}}$, $i \geq 0$, $0 \leq j \leq J_0$, and the following estimates hold at each interaction diamond Δ_{ij}, $i \geq 0, j \geq J_0 - 1$:

$$\|\mathbf{z}_{ij} - \mathbf{z}_{i,j-1}\| \leq \sum_{\substack{l=i,i+1 \\ p=1,2}} \left\{ |\gamma_{l,j-1}^p| + H\left(|\gamma_{l,j-1}^p|\right) \right\} + G_1 \Delta t, \tag{7.33}$$

$$|z_{ij} - z_{i,j-1}| \leq \sum_{\substack{l=i,i+1 \\ p=1,2}} |\gamma_{l,j-1}^p| + G_1 \Delta t, \tag{7.34}$$

$$|\gamma_{ij}^{OUT}| - |\gamma_{ij}^{IN}| \leq \left\{ |\gamma_{ij}^{IN}| + H\left(|\gamma_{ij}^{IN}|\right) \right\} e^{G_1 \Delta t} G_1 \Delta t, \tag{7.35}$$

where,

$$G_1 \equiv G_1(\bar{B},\bar{M},\bar{S}) = 2\sqrt{2}\frac{G_0(\bar{B},\bar{M})}{r_0}(\kappa c^2 r_0 \bar{S} + 1), \tag{7.36}$$

$$G_0 \equiv G_0(\bar{B},\bar{M}) = K_1\sqrt{G_{AB}(\bar{B},\bar{M})}, \tag{7.37}$$

$$K_1 = \frac{8c^4}{(c^2-\sigma^2)^2}, \tag{7.38}$$

$$G_{AB} = A_{r_0} B_{r_0} \exp\left\{\frac{8\bar{B}\bar{M}}{r_0}\right\}. \tag{7.39}$$

Moreover,

$$1 \leq B_{\Delta x}(x, t_j), \tag{7.40}$$
$$0 < A_{r_0} < A_{\Delta x}(x, t_j) \leq G_{AB}, \tag{7.41}$$
$$0 < x u^0_{\Delta x}(x, t_j) \leq \frac{c^2 + \sigma^2 \bar{v}^2}{c^2 - \bar{v}^2} \bar{S}, \tag{7.42}$$

and

$$|A'_{\Delta x}(x, t_j)| \leq \left(\frac{1}{r_0} + \kappa \frac{c^2 + \sigma^2 \bar{v}^2}{c^2 - \bar{v}^2} \bar{S}\right) \bar{B} G_{AB}, \tag{7.43}$$

$$|B'_{\Delta x}(x, t_j)| \leq \left(\frac{1}{r_0} + \kappa \frac{c^2 + \sigma^2 \bar{v}^2}{c^2 - \bar{v}^2} \bar{S}\right) \bar{B}^2. \tag{7.44}$$

for all $x \geq r_0$, $0 \leq t_j \leq T_0 = t_{J_0} \leq 1$.

Note, again, that (7.43) gives the Lipschitz continuity in x of $A_{\Delta x}$ and $B_{\Delta x}$ and (7.42) implies that $\rho_{\Delta x} > 0$ for $0 \leq t \leq T_0$.

Proof: This follows directly from (7.24), (7.26) and (7.27), together with Propositions 2 and 10. □

Corollary 1 *Assume that the approximate solution $u_{\Delta x}, \mathbf{A}_{\Delta x}$ satisfies the conditions (7.28)-(7.32) of Theorem 4 up to some time T_0, $0 < T_0 = t_{J_0} \leq 1$, and assume further that there exists constants L, V_0 such that*

$$\sum_{i_1 \leq i \leq i_2,\, p=1,2} |\gamma^p_{i0}| < V_0, \tag{7.45}$$

for all $|x_{i_2} - x_{i_1}| \leq L$. Then for any constant $\alpha > 1$:

(A) The following total variation bound holds:

$$\sum_{i_1 \leq i \leq i_2,\, p=1,2} |\gamma^p_{ij}| \leq \alpha \left(1 + \frac{4 t_j \sqrt{G_{AB}}}{L}\right) V_0 \leq \alpha \bar{V}_*, \tag{7.46}$$

for all $|x_{i_2} - x_{i_1}| \leq L$, so long as $t_j \leq \mathrm{Min}\{T_\alpha, T_0\} \leq 1$ where

$$T_\alpha = \left(\frac{1}{G_1 e^{G_1}}\right) \frac{(\alpha - 1) \bar{V}_*}{\{\alpha \bar{V}_* + H(\alpha \bar{V}_*)\}}, \tag{7.47}$$

$$\bar{V}_* \equiv \left(1 + \frac{4 \sqrt{G_{AB}}}{L}\right) V_0. \tag{7.48}$$

(B) The following L^1_{loc} bounds hold:

$$\int_{x_{i_1}}^{x_{i_2}} \|\mathbf{z}_{\Delta x}(x,t_{j_2}) - \mathbf{z}_{\Delta x}(x,t_{j_1})\| \, dx$$
$$\leq \left\{ 4\sqrt{G_{AB}} \left[\alpha \bar{V}_* + H(\alpha \bar{V}_*)\right] + G_1 |x_{i_2} - x_{i_1}| \right\} |t_{j_2} - t_{j_1}|, \quad (7.49)$$

$$\int_{x_{i_1}}^{x_{i_2}} |z_{\Delta x}(x,t_{j_2}) - z_{\Delta x}(x,t_{j_1})| \, dx$$
$$\leq \left\{ 4\sqrt{G_{AB}} \left[\alpha \bar{V}_*\right] + G_1 |x_{i_2} - x_{i_1}| \right\} |t_{j_2} - t_{j_1}|, \quad (7.50)$$

for any $t_{j_1} \leq t_{j_2} \leq Min\{T_\alpha, T_0\} \leq 1$, and any $r_0 \leq x_{i_1} \leq x_{i_2} < \infty$.

(C) The following bounds on the supnorm hold:

$$|z_{ij} - z_{i+j,0}| \leq \alpha \left(1 + \frac{4t_j \sqrt{G_{AB}}}{L}\right) V_0 + 2\sqrt{G_{AB}} G_1 t_j, \quad (7.51)$$

$$|w_{ij} - w_{i+j,0}| \leq H\left(\alpha \left(1 + \frac{4t_j \sqrt{G_{AB}}}{L}\right) V_0\right) + 2\sqrt{G_{AB}} G_1 t_j, \quad (7.52)$$

$$\|\mathbf{z}_{ij} - \mathbf{z}_{i+j,0}\| \leq \alpha \left(1 + \frac{4t_j \sqrt{G_{AB}}}{L}\right) V_0 \quad (7.53)$$
$$+ H\left(\alpha \left(1 + \frac{4t_j \sqrt{G_{AB}}}{L}\right) V_0\right) + 2\sqrt{G_{AB}} G_1 t_j,$$

for all $t_j \leq Min(T_\alpha, T_0) \leq 1$.

The motivation for choosing the factor $\left(1 + \frac{4t_j \sqrt{G_{AB}}}{L}\right)$ in (7.85) of (A) is that since

$$\frac{\Delta x}{\Delta t} = \Lambda = 2\sqrt{G_{AB}(\bar{B}, \bar{M})}, \quad (7.54)$$

it follows that

$$\left(1 + \frac{4t_j \sqrt{G_{AB}}}{L}\right) \geq \frac{x_{i_2} - x_{i_1} + 4\frac{\Delta x}{\Delta t} t_j}{L}, \quad (7.55)$$

where the RHS of (7.55) dominates the number of intervals of length L contained within the domain of dependence of $[x_{i_1}, x_{i_2}]$ at time level t_j. Note that the appearance of $t_j \sqrt{G_{AB}}$ in $\left(1 + \frac{4t_j \sqrt{G_{AB}}}{L}\right)$ (7.46) is important because the LHS of (7.46) can be estimated independently of $\bar{M}, \bar{B}, \bar{S}$ and $\bar{v}$ for t_j sufficiently small.

Regarding part (C), note that $\mathbf{z}_{i+j,0} = \mathbf{z}_{\Delta x}(x_i + t_j \frac{\Delta x}{\Delta t}, 0)$ where $t_j \frac{\Delta x}{\Delta t} = 2t_j \sqrt{G_{AB}}$ depends on $\bar{M}, \bar{B}$. Note also that since

$$|v_{\Delta x}(x, t_j)| \leq \bar{v}_j < c \;\; iff \;\; |w_{\Delta x}(x, t_j)| \leq \bar{w}_j = \left|\ln \frac{c + \bar{v}_j}{c - \bar{v}_j}\right|, \qquad (7.56)$$

it follows from (7.52) that if initially $|w_{\Delta x}(x_i, 0)| \leq \bar{w}_0$, then

$$|w_{\Delta x}(x_i, t_j)| \leq \bar{w}_0 + H\left(\alpha\left(1 + \frac{4t_j\sqrt{G_{AB}}}{L}\right)V_0\right) + G_1 t_j \sqrt{G_{AB}} = \bar{w}_j, \quad (7.57)$$

for all $x_i \geq r_0$, $t_j \leq Min\{T_\alpha, T_0\} \leq 1$. Thus $|w|$ is bounded uniformly and v is bounded uniformly away from c at each $t_j \leq Min\{T_\alpha, T_0\} \leq 1$ so long as these bounds hold initially.

Proof of (A): Assume $0 < t_j \leq T_0$, and consider the interaction diamonds Δ_{ij}, $i = i_1, \ldots, i_2$. Then by (7.35), (if $i_1 > 0$),

$$\sum_{i_1 \leq i \leq i_2, \, p=1,2} |\gamma_{ij}^p| = \sum_{i_1 \leq i \leq i_2} |\gamma_{ij}^{OUT}| \qquad (7.58)$$

$$\leq \sum_{i_1 \leq i \leq i_2} |\gamma_{ij}^{IN}| + \sum_{i_1 \leq i \leq i_2} \left\{|\gamma_{ij}^{IN}| + H\left(|\gamma_{ij}^{IN}|\right)\right\} e^{G_1 \Delta t} G_1 \Delta t$$

$$\leq \sum_{i_1-1 \leq i \leq i_2+1} |\gamma_{i,j-1}| + \sum_{i_1-1 \leq i \leq i_2+1} \left\{|\gamma_{i,j-1}| + H\left(|\gamma_{i,j-1}|\right)\right\} e^{G_1 \Delta t} G_1 \Delta t.$$

More generally, let

$$V_j = \sum_{i_1 \leq i \leq i_2} |\gamma_{ij}^{OUT}|, \qquad (7.59)$$

$$V_{j-1} = \sum_{\partial(i_1-1) \leq i \leq i_2+1} |\gamma_{i,j-1}^{OUT}|, \qquad (7.60)$$

$$V_0 = \sum_{\partial(i_1-j) \leq i \leq i_2+j} |\gamma_{i,0}^{OUT}|, \qquad (7.61)$$

where to account for the boundary at $r = r_0$, we let

$$\partial(i_1 - j) = \begin{cases} 0 & i_1 - j \leq 0 \\ i_1 - j & otherwise \end{cases}. \qquad (7.62)$$

Then (7.58) together with the convexity of H imply that

$$V_k - V_{k-1} \leq \{V_{k-1} + H(V_{k-1})\} e^{G_1 \Delta t} G_1 \Delta t, \qquad (7.63)$$

for all $k \leq j$. To estimate V_j, define

$$\bar{V}_0 = \left\{1 + \frac{x_{i_2+j} - x_{\partial(i_1-j)}}{L}\right\} V_0 \geq \sum_{\partial(i_1-j) \leq i \leq i_2+j,\ p=1,2} |\gamma_{i0}^p|, \qquad (7.64)$$

c.f. (7.45), and inductively let

$$\bar{V}_k = \bar{V}_{k-1} + \{\bar{V}_{k-1} + H(\bar{V}_{k-1})\} e^{G_1 \Delta t} G_1 \Delta t, \qquad (7.65)$$

to define $\bar{V}_k$ for all $k \leq j$. Note that

$$\bar{V}_* \equiv \left(1 + \frac{4\sqrt{G_{AB}}}{L}\right) V_0 \geq \left(1 + \frac{4 t_j \sqrt{G_{AB}}}{L}\right) V_0 \geq \bar{V}_0, \qquad (7.66)$$

where we use that

$$\left\{1 + \frac{|x_{i_2+j} - x_{\partial(i_1-j)}|}{L}\right\} \leq \left\{\frac{L + 2 t_j \frac{\Delta x}{\Delta t}}{L}\right\} \qquad (7.67)$$

dominates the number of intervals of length L contained in $|x_{i_2+j} - x_{\partial(i_1-j)}|$. Now $\bar{V}_k$ increases with k, and by induction using (7.63) it follows that

$$\bar{V}_k \geq V_k, \qquad (7.68)$$

for all $k \leq j$. Thus to estimate V_j, it suffices to estimate $\bar{V}_j$. To this end, fix $\alpha > 1$, and let T_α be given by (7.47).

Claim: $\bar{V}_j \leq \alpha \bar{V}_0$ for all $t_j \leq Min\{T_\alpha, T_0\} \leq 1$.

To prove the claim, assume that $t_j \leq Min\{T_\alpha, T_0\} \leq 1$, and t_{j+1} is the first time such that

$$\bar{V}_{j+1} > \alpha \bar{V}_0. \qquad (7.69)$$

Then for $t_k \leq t_j$,

$$\bar{V}_k - \bar{V}_{k-1} \leq \{\alpha \bar{V}_0 + H(\alpha \bar{V}_0)\} e^{G_1 \Delta t} G_1 \Delta t, \qquad (7.70)$$

and summing we obtain

$$\bar{V}_k - \bar{V}_0 \leq \{\alpha \bar{V}_0 + H(\alpha \bar{V}_0)\} e^{G_1 \Delta t} G_1 t_j. \qquad (7.71)$$

But solving for t_j in (7.71) shows that

$$\{\alpha \bar{V}_0 + H(\alpha \bar{V}_0)\} e^{G_1 \Delta t} G_1 t_j \leq (\alpha - 1) \bar{V}_0, \tag{7.72}$$

so long as

$$t_j \leq \left(\frac{1}{G_1 e^{G_1}}\right) \frac{(\alpha - 1)\bar{V}_0}{\{\alpha \bar{V}_0 + H(\alpha \bar{V}_0)\}}.$$

But

$$T_\alpha \leq \left(\frac{1}{G_1 e^{G_1}}\right) \frac{(\alpha - 1)\bar{V}_0}{\{\alpha \bar{V}_0 + H(\alpha \bar{V}_0)\}}, \tag{7.73}$$

and so it follows (inductively) from (7.73) that the bound (7.69) is maintained so long as $t_j \leq Min\{T_\alpha, T_0\} \leq 1$, as claimed.

In light of (7.66), it follows that

$$\sum_{i_1 \leq i \leq i_2} |\gamma_{ij}^p| = V_j \leq \bar{V}_j \leq \alpha \bar{V}_0 < \alpha \left(1 + \frac{4t_j \sqrt{G_{AB}}}{L}\right) V_0 \leq \alpha \bar{V}_*$$

for all $t_j \leq Min\{T_\alpha, T_0\} \leq 1$, which is (7.46). The proof of (A) is complete.

Proof of (B): For (7.49), estimate as follows:

$$\int_{x_{i_1}}^{x_{i_2}} \|\mathbf{z}_{\Delta x}(x, t_{j_2}) - \mathbf{z}_{\Delta x}(x, t_{j_1})\| \, dx$$

$$= \sum_{i=i_1}^{i_2-1} \|\mathbf{z}_{ij_2} - \mathbf{z}_{ij_1}\| \Delta x \leq \sum_{i=i_1}^{i_2-1} \sum_{j=j_1+1}^{j_2} \|\mathbf{z}_{ij} - \mathbf{z}_{i,j-1}\| \Delta x$$

$$\leq \sum_{i=i_1}^{i_2-1} \sum_{j=j_1+1}^{j_2} \left\{ \sum_{\substack{l=i,i+1 \\ p=1,2}} \left[|\gamma_{i,j-1}^p| + H\left(|\gamma_{i,j-1}^p|\right)\right] + G_1 \Delta t \right\} \Delta x$$

$$\leq 2 \sum_{j=j_1+1}^{j_2} \left[\sum_{i=i_1}^{i_2} \sum_{p=1,2} \left\{|\gamma_{i,j-1}^p| + H\left(|\gamma_{i,j-1}^p|\right)\right\}\right] \Delta x + G_1 |x_{i_2} - x_{i_1}| |t_{j_2} - t_{j_2}|$$

$$\leq \left\{2 \left[\alpha \bar{V}_* + H(\alpha \bar{V}_*)\right] \frac{\Delta x}{\Delta t} + G_1 |x_{i_2} - x_{i_1}|\right\} |t_{j_2} - t_{j_2}|$$

where we have used (7.33) and (7.46). In light of (7.28), this verifies (7.49). Inequality (7.50) follows by the same argument using (7.34) in place of (7.33).

Proof of (C): Note first that (7.33) and (7.34) directly imply that

$$\|\mathbf{z}_{ij} - \mathbf{z}_{i0}\| \leq \sum_{\substack{l = i, i+1 \\ 0 \leq k \leq j-1 \\ p = 1, 2}} \{|\gamma_{lk}^p| + H(|\gamma_{lk}^p|)\} + G_1 t_j, \quad (7.74)$$

$$|z_{ij} - z_{i0}| \leq \sum_{\substack{l = i, i+1 \\ 0 \leq k \leq j-1 \\ p = 1, 2}} |\gamma_{lk}^p| + G_1 t_j. \quad (7.75)$$

Unfortunately, we cannot use (7.74) directly to estimate $\|\mathbf{z}_{ij} - \mathbf{z}_{i0}\|$ because we cannot bound the right-hand-side by V_0 without introducing wave-tracing to identify waves at time t_j with waves at time $t = 0$. To get around this, we estimate $\|\mathbf{z}_{ij} - \mathbf{z}_{i0}\|$ as follows.

Let (x_i, t_j) be fixed. Let J_{ij}^R denote the piecewise linear I-curve that connects mesh points $(x_i, t_{j+\frac{1}{2}})$ to $(x_i + a_j\Delta x_j, t_j)$ to $(x_{i+1}, t_{(j-1)+\frac{1}{2}})$ and so on, continuing downward and to the right until you reach $(x_{i+j} + a_0\Delta x, 0)$ at time $t_0 = 0$. Let J_{ij}^L connect $(x_i, t_{j+\frac{1}{2}})$ to $(x_{i-1} + a_j\Delta x, t_j)$ to $(x_{i-1}, t_{(j-1)+\frac{1}{2}})$ and so on, continuing downward and to the left until one reaches $t = 0$ at (x_{i-j-1}, t_0) or else stop at $r = r_0$ at the point $(r_0, t_{j_0+\frac{1}{2}})$, (see Figure 6). Let J_{ij} denote the I-curve $J_{ij} = J_{ij}^L \cup J_{ij}^R$, and recall from [8, 23, 24], that one can connect J_{ij} by a sequence of I-curves, $J_0, \ldots, J_N = J_{ij}$ such that J_{k+1} is an immediate successor of J_k, and J_0 is the I-curve that crosses the waves γ_{i0}^p between $i = \partial(i - i_1)$ and $i = i + j$. (See figure 4.) Since J_k differs from J_{k+1} by a single interaction diamond, it follows by induction using (7.58), and the argument (7.58)-(7.73), that

$$\sum_{J_{ij}} |\gamma_{ij}^p| \leq \alpha \left(1 + \frac{4t_j\sqrt{G_{AB}}}{L}\right) V_0, \quad (7.76)$$

where $\sum_{J_{ij}} |\gamma_{ij}^p|$ denotes the sum of the waves γ_{ij}^p that cross the curve J_{ij}. (We have used the assumption $t_j \leq Min\{T_\alpha, T_0\} \leq 1$.) From this it follows that

$$\sum_{J_{ij}^R} |\gamma_{ij}^p| \leq \alpha \left(1 + \frac{4t_j\sqrt{G_{AB}}}{L}\right) V_0. \quad (7.77)$$

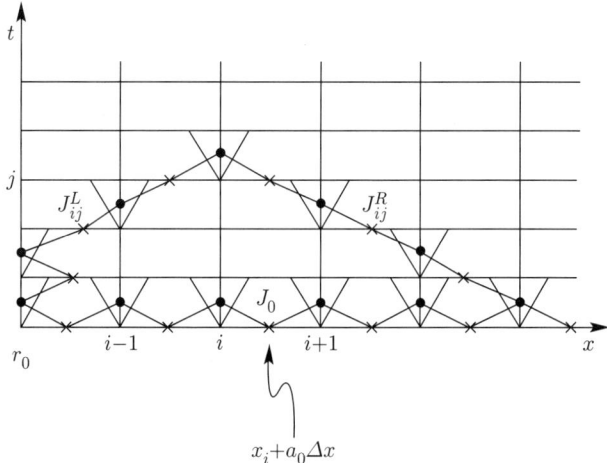

Figure 6: The I-curves J_0, J_{ij}^L and J_{ij}^R

But $\sum_{J_{ij}^R} |\gamma_{ij}^p|$ bounds the total variation in z between the state $\mathbf{z}_{i+j,0}$ and the state $\mathbf{z}_{ij}$, except for the change in z that occurs between $\mathbf{z}_{i'j'}^{RP}$ and $\mathbf{z}_{i'j'}$ at each $(x_{i'}, t_{j'})$ that lies on the I-curve J_{ij}^R. But by (7.25), we know that

$$\|z_{i'j'}^{RP} - z_{i'j'}\| \leq G_1 \Delta x, \qquad (7.78)$$

so it follows that

$$|z_{ij} - z_{i+j,0}| \leq \alpha \left(1 + \frac{4t_j \sqrt{G_{AB}}}{L}\right) V_0 + G_1 t_j \frac{\Delta x}{\Delta t}. \qquad (7.79)$$

which verifies (7.51) in light of (7.54), (7.54). Also, since

$$\|\gamma_{ij}^p\| \leq |\gamma_{ij}^p| + H\left(|\gamma_{ij}^p|\right), \qquad (7.80)$$

where $H\left(|\gamma_{ij}^p|\right)$ bounds the change in w across wave γ_{ij}^p, it follows that

$$|w_{ij} - w_{i+j,0}| \leq \sum_{J_{ij}^R} H\left(|\gamma_{ij}^p|\right) + G_1 t_j \frac{\Delta x}{\Delta t},$$

$$\|\mathbf{z}_{ij} - \mathbf{z}_{i+j,0}\| \leq \sum_{J_{ij}^R} |\gamma_{ij}^p| + H\left(|\gamma_{ij}^p|\right) + G_1 t_j \frac{\Delta x}{\Delta t},$$

and so using (7.77), (which again uses $t_j \leq Min\{T_\alpha, T_0\} \leq 1$), we obtain (7.52) and (7.53). This completes the proof of Corollary 1. $\square$

SHOCK-WAVE SOLUTIONS OF THE EINSTEIN EQUATIONS   59

In order to unify the estimates in (B) and (C), assume that $|x_{i_2} - x_{i_1}| \leq L$, and set $G_2 \equiv G_2(\bar{B}, \bar{M}, \bar{S})$ equal to

$$G_2 = Max\left\{4\sqrt{G_{AB}}\left[\alpha \bar{V}_* + H(\alpha \bar{V}_*)\right] + G_1 L, 2\sqrt{G_{AB}} G_1, G_1 e^{G_1}, \sqrt{G_{AB}}\right\}, \tag{7.81}$$

where $G_1 \equiv G_1(\bar{B}, \bar{M}, \bar{S})$, $G_{AB} \equiv G_{AB}(\bar{B}, \bar{M})$ and $\bar{V}_*$ are defined in (7.36), (7.39) and (7.48), respectively. (Note that $\bar{V}_*$, and hence G_2, also depend on V_0, but for our purposes we only keep track of the dependence on $\bar{M}, \bar{B}, \bar{S}, \bar{v}$, the constants that are not yet determined by the initial data.) Then the following corollary is a simplification of Corollary 1.

Corollary 2 *Assume that the approximate solution $u_{\Delta x}, \mathbf{A}_{\Delta x}$ satisfies the conditions (7.28)-(7.32) of Theorem 4 up to some time T_0, $0 < T_0 = t_{J_0} \leq 1$, and assume that there exists constants L, V_0 such that*

$$\sum_{i_1 \leq i \leq i_2,\, p=1,2} |\gamma_{i0}^p| < V_0, \tag{7.82}$$

for all $|x_{i_2} - x_{i_1}| \leq L$, and assume that $\alpha = 2$, c.f. (7.46). Then:

(A) The following total variation bound holds:

$$\sum_{i_1 \leq i \leq i_2,\, p=1,2} |\gamma_{ij}^p| < 2\bar{V}_*, \tag{7.83}$$

for all $|x_{i_2} - x_{i_1}| \leq L$, so long as $t_j \leq Min\{T_2, T_0\} \leq 1$, where

$$T_2 = \left(\frac{1}{G_2}\right) \frac{\bar{V}_*}{\left\{2\bar{V}_* + H(2\bar{V}_*)\right\}}, \tag{7.84}$$

$$\bar{V}_* = \left(1 + \frac{4\sqrt{G_{AB}}}{L}\right) V_0. \tag{7.85}$$

(B) The following L^1_{loc} bounds hold:

$$\int_{x_{i_1}}^{x_{i_2}} \|\mathbf{z}_{\Delta x}(x, t_{j_2}) - \mathbf{z}_{\Delta x}(x, t_{j_1})\| \, dx \leq G_2 |t_{j_2} - t_{j_1}|, \tag{7.86}$$

and

$$\int_{x_{i_1}}^{x_{i_2}} |z_{\Delta x}(x, t_{j_2}) - z_{\Delta x}(x, t_{j_1})| \, dx \leq G_2 |t_{j_2} - t_{j_1}|, \tag{7.87}$$

for all $r_0 \leq x_{i_1} < x_{i_2} < \infty$, $|x_{i_2} - x_{i_1}| \leq L$, $t_j \leq Min\{T_0, T_2\}$.
(C) The following bounds on the supnorm hold.

$$|z_{ij} - z_{i+j,0}| \leq F_0^*(G_2 \cdot t_j), \qquad (7.88)$$
$$|w_{ij} - w_{i+j,0}| \leq F_0^*(G_2 \cdot t_j), \qquad (7.89)$$
$$\|\mathbf{z}_{ij} - \mathbf{z}_{i+j,0}\| \leq F_0^*(G_2 \cdot t_j), \qquad (7.90)$$

for all $x_i \geq r_0$, $t_j \leq \text{Min}\{T_0, T_2\}$, where

$$F_0^*(\xi) = 2\left(1 + \frac{4\xi}{L}\right)V_0 + H\left(2\left(1 + \frac{4\xi}{L}\right)V_0\right) + \xi. \qquad (7.91)$$

Again, consistent with our notation, the functions $G_2(\cdot, \cdot, \cdot)$ and $F_0^*(\xi)$ depend only on constants σ, c, K_0, r_0, L and V_0 that depend only on the initial data, and so the functions $G_2(\cdot, \cdot, \cdot)$ and $F_0^*(\xi)$ are independent of the constants $\bar{M}, \bar{B}, \bar{S}, \bar{v}$. The functions $G_2(\cdot, \cdot, \cdot)$ and $F_0^*(\xi)$ are also increasing functions of each argument. The main point is that constants that depend on $\bar{M}, \bar{B}, \bar{S}$ or $\bar{v}$ in the estimates (7.86)-(7.90), are organized into the single constant G_2, (which happens to be independent of $\bar{v}$), and which is always multiplied by the factor t_j. Thus estimates independent of $\bar{M}, \bar{B}, \bar{S}$ and $\bar{v}$ can obtained by making t_j sufficiently small. Note that the formula for $F_0^*(\xi)$ is obtained by substituting 2 for α, and ξ for $t_j\sqrt{G_{AB}}$ and $2\sqrt{G_{AB}}G_1 t_j$, on the RHS of (7.53).

8 The Elimination of Assumptions

In this section we show that the assumptions (7.29)-(7.32) in Corollary 2, Theorem 4 above, needn't be assumed, but are implied by values of $\bar{M}, \bar{B}, \bar{S}, \bar{v}$ that can be defined in terms of the initial data alone, subject to restrictions on the time T_0. Once we succeed with this replacement, Theorem 4 and Corollary 2 provide the uniform bounds required to apply the Oleinik compactness argument demonstrating the compactness of approximate solutons up to some finite time T. To start, consider first the bound (7.32) for v. Since the constant G_1 in Corollary 2 is independent v, it follows that we can achieve (7.32) for a value of $\bar{v}$ defined in terms of the bound on v at time $t = 0$. Indeed, assume that the initial data $v_{\Delta x}(x, 0)$ satisfies

$$|v_{\Delta x}(x, 0)| \leq \bar{v}_0 < c \iff |w_{\Delta x}(x, 0)| \leq \left|\ln\left(\frac{c + \bar{v}_0}{c - \bar{v}_0}\right)\right| \equiv \bar{w}_0, \qquad (8.1)$$

for all $r_0 \leq x$. Then assuming the hypotheses of Corollary 2, it follows from (7.89) that

$$|w_{\Delta x}(x, t_j)| \leq \bar{w}_0 + F_0^*(G_2 t_j), \tag{8.2}$$

for all $r_0 \leq x$, $t_j \leq Min\{T_2, T_0\}$. Therefore, if we *define* $\bar{v}$ so that $\bar{w} = \left|\ln\left(\frac{c+\bar{v}}{c-\bar{v}}\right)\right|$, where

$$\bar{w} \equiv \bar{w}_0 + F_0^*(G_2 t_j), \tag{8.3}$$

then (7.32) is a *consequence* of our other assumptions. Indeed, to make a rigorous proof out of this, just define $\bar{v}$ by (8.3), (8.2), and let T_v be the first time at which $|v| \leq \bar{v}$ fails. The argument that leads to the choice of $\bar{w}$ in (8.2) then shows that $T_v \geq Min\{T_2, T_0\}$.

Similarly, we now use (7.88) to show that $\bar{S}$ can be defined in terms of an initial bound $\bar{S}_0$ on $S_{\Delta x}(x, 0)$ in such a way that (7.31) can be eliminated as an assumption in Corollary 2 because it follows as a consequence of our other assumptions. In this case, however, (as in the case of $\bar{M}$ and $\bar{B}$), the constant G_1 depends on $\bar{S}$, so we need a corresponding restriction $t \leq T_S$ for some $T_S \ll 1$. Indeed, assume that the initial data $S_{\Delta x}(x, 0)$ satisfies

$$0 < S_{\Delta x}(x, 0) \leq \bar{S}_0, \tag{8.4}$$

for all $r_0 \leq x$. Then assuming the hypotheses of Corollary 2, it follows from (7.88) that

$$K_0 \ln \rho_{ij} - K_0 \ln \rho_{i+j,0} \leq F_0^* (G_2 \cdot t_j), \tag{8.5}$$

and so

$$0 < \rho_{ij} \leq F_1^* (G_2 \cdot t_j) \rho_{i+j,0} \tag{8.6}$$

where

$$F_1^*(\xi) = \exp\left\{\frac{F_0(\xi)}{K_0}\right\} \geq 1. \tag{8.7}$$

It follows from (8.4) that

$$S_{ij} = x_i \rho_{ij} \leq F_1^* (G_2 \cdot t_j) x_i \rho_{i+j,0} \leq F_1^* (G_2 \cdot t_j) \bar{S}_0. \tag{8.8}$$

Inequality (8.8) tells us that if we choose $\bar{S} \geq F_1^* (0) \bar{S}_0$, say choose

$$\bar{S} = 2F_1^*(0)\bar{S}_0, \tag{8.9}$$

and set

$$T_S = Sup\{t : F_1^* (G_2 \cdot t_j) \leq 2F_1^*(0), \text{ for all } t_j \leq t\}, \tag{8.10}$$

then assumption (7.29) of Corollary 2, Theorem 4, (that $0 < S_{\Delta x}(x,t) \leq \bar{S}$), can be replaced by the condition that $\bar{S}$ is defined in (8.9), together with the assumption that $t_j \leq T_S$, where T_S is defined in (8.9). Note that this argument relies on the fact that the function $F_1^*(\cdot)$ is independent of $\bar{S}$.

We now derive formulas analogous to (8.9), (8.10), for $\bar{M}, T_M$ and $\bar{B}, T_B$, so that assumptions (7.29) and (7.30) of Corollary 2, Theorem 4, can be replaced by the condition that $\bar{M}, \bar{B}$ be defined by the values given in the formulas, together with $t_j \leq T_M$ and $t_j \leq T_B$, respectively. So consider next the total mass

$$M_{\Delta x}(\infty, t_j) = M_{r_0} + M_j, \quad M_j \equiv \frac{\kappa}{2} \int_{r_0}^{\infty} u_{\Delta x}^0(r, t_j) r^2 \, dr. \tag{8.11}$$

Using $u^0 = T_M^{00}$ and $\ln w = \frac{c+v}{c-v}$ in (1.15), we obtain

$$u^0 = \frac{1}{2} \left\{ (1+\sigma^2) \cosh w + (1-\sigma) \right\} \rho. \tag{8.12}$$

Thus it follows from (8.3) and (8.6) that

$$u_{\Delta x}^0(x_i, t_j) \leq \frac{1}{2} \left\{ (1+\sigma^2) \cosh \bar{w} + (1-\sigma) \right\} F_1^*(G_2 \cdot t_j) \rho_{\Delta x}(x_{i+j}, 0). \tag{8.13}$$

Using this in (8.11) we obtain

$$\begin{aligned} M_j &\leq M_{r_0} + F_2^*(G_2 \cdot t_j) \frac{\kappa}{2} \int_{r_0}^{\infty} \rho_{\Delta x}(r + j\Delta x, 0) r^2 \, dr \\ &\leq M_{r_0} + F_2^*(G_2 \cdot t_j) \frac{\kappa}{2} \int_{r_0}^{\infty} u_{\Delta x}^0(r, 0) r^2 \, dr \\ &= M_{r_0} + F_2^*(G_2 \cdot t_j) M_0, \end{aligned} \tag{8.14}$$

where

$$F_2^*(\xi) \equiv \frac{1}{2} \left\{ (1+\sigma^2) \cosh(\bar{w}) + (1-\sigma) \right\} F_1^*(\xi), \tag{8.15}$$

$$M_0 = \frac{\kappa}{2} \int_{r_0}^{\infty} u_{\Delta x}^0(r, 0) r^2 \, dr. \tag{8.16}$$

Inequality (8.14) tells us that if we choose $\bar{M} \geq M_{r_0} + F_2^*(0) M_0$, say choose

$$\bar{M} = M_{r_0} + 2 F_2^*(0) M_0, \tag{8.17}$$

and set

$$T_M = Sup\{t : M_{r_0} + F_2^*(G_2 \cdot t_j) M_0 \leq \bar{M}, \text{ for all } t_j \leq t\}. \tag{8.18}$$

then assumption (7.29) of Corollary 2, Theorem 4, can be replaced by the condition that $\bar{M}$ is defined in (8.17), together with the assumption that $t_j \leq T_M$, where T_M is defined in (8.18).

We now turn to the problem of defining $\bar{B}, T_B$ so as to replace the final assumption (7.30) of Theorem 3. Since

$$B_{\Delta x}(x,t) = \frac{1}{1 - \frac{2GM_{\Delta x}(x,t)}{x}}, \tag{8.19}$$

it follows that to accomplish this, we must estimate the change in $M_{\Delta x}(x, t_j)$ between times $t = 0$ and $t = t_j$, assuming that Corollary 2, Theorem 4, applies. More generally, assume that (7.29)-(7.32), and hence Theorem 4, hold up to time t_j, and assume that $0 \leq t_{j_0} < t_j$. We estimate

$$|M_{\Delta x}(x, t_j) - M_{\Delta x}(x, t_{j_0})| \leq \frac{\kappa}{2} \int_{r_0}^{x} |u_{\Delta x}^0(r, t_j) - u_{\Delta x}^0(r, t_{j_0})| r^2 \, dr. \tag{8.20}$$

To start, let

$$\begin{aligned}
\Delta u^0 &= u_{\Delta x}^0(r, t_j) - u_{\Delta x}^0(r, t_{j_0}), \\
\Delta w &= w_{\Delta x}(r, t_j) - w_{\Delta x}(r, t_{j_0}), \\
\Delta \rho &= \rho_{\Delta x}^0(r, t_j) - \rho_{\Delta x}^0(r, t_{j_0}),
\end{aligned} \tag{8.21}$$

etc. Then

$$|\Delta u^0| \leq \left\|\frac{\partial u^0}{\partial w}\right\|_\infty |\Delta w| + \left\|\frac{\partial u^0}{\partial z}\right\|_\infty |\Delta z|. \tag{8.22}$$

From (8.12) we calculate

$$\frac{\partial u^0}{\partial z} = \frac{1}{2K_0}\{(1+\sigma^2)\cosh w + (1-\sigma)\}\rho, \tag{8.23}$$

$$\frac{\partial u^0}{\partial w} = \frac{1+\sigma^2}{2}(\sinh w)\rho. \tag{8.24}$$

Since $\left|\frac{\partial u^0}{\partial w}\right| \leq \frac{\partial u^0}{\partial z}$, it follows from (8.22) that

$$|\Delta u^0| \leq \frac{1}{K_0}\frac{1}{2}\left\{(1+\sigma^2)\cosh(\bar{w}) + (1-\sigma)\right\}\|\rho\|_\infty \{|\Delta w| + |\Delta z|\}$$
$$\leq \frac{\sqrt{2}}{K_0}\frac{F_2^*(G_2 \cdot t_j)}{F_1^*(G_2 \cdot t_j)}\|\rho\|_\infty\|\Delta\mathbf{z}\|. \tag{8.25}$$

Putting (8.25) into (8.20) and using

$$\rho_{\Delta x}(x,t) \leq F_1^*(G_2 \cdot t_j)\,\rho_{\Delta x}(x+j\Delta x, 0), \tag{8.26}$$

we obtain

$$|M_{\Delta x}(x,t_j) - M_{\Delta x}(x,t_{j_0})| \leq \frac{\kappa F_2^*(G_2 \cdot t_j)}{\sqrt{2}K_0}\int_{r_0}^x \rho_{\Delta x}(r+j\Delta x, 0)\,\|\Delta\mathbf{z}\|r^2\,dr. \tag{8.27}$$

We use (8.27) again below, but for now we can continue from (8.27) to obtain

$$\begin{aligned}\frac{\kappa F_2^*(G_2 \cdot t_j)}{\sqrt{2}K_0}\int_{r_0}^x \rho_{\Delta x}(r+j\Delta x, 0)\,\|\Delta\mathbf{z}\|r^2\,dr \\ \leq \frac{\kappa F_2^*(G_2 \cdot t_j)\bar{S}x}{\sqrt{2}K_0}\int_{r_0}^x \|\Delta\mathbf{z}\|\,dr \\ \leq \frac{\kappa F_2^*(G_2 \cdot t_j)\bar{S}x^2}{\sqrt{2}K_0 L}G_2|t_j - t_{j_0}|,\end{aligned} \tag{8.28}$$

where we have used

$$\int_{r_0}^x \|\Delta\mathbf{z}\|\,dr \leq \frac{x}{L}G_2|t_{j_2} - t_{j_1}|, \tag{8.29}$$

a consequence of (7.86). Note that the factor x/L bounds the number of intervals of length L between r_0 and x. We record this as a Corollary of Theorem 4:

Corollary 3 *Assume Corollary 2, Theorem 4, applies up to time T_0. Then*

$$|M_{\Delta x}(x,t_j) - M_{\Delta x}(x,t_{j_0})| \leq \frac{\kappa F_2^*(G_2 \cdot t_j)\bar{S}x^2}{\sqrt{2}K_0 L}G_2|t_j - t_{j_0}|, \tag{8.30}$$

for all $0 \leq t_{j_0} \leq t_j \leq T_0$.

In particular, ignoring errors of order Δx, (8.30) implies the local Lipschitz in time continuity of $M_{\Delta x}$, (and hence of $B_{\Delta x}$ and $A_{\Delta x}$).

We can estimate $|M_{\Delta x}(x, t_j) - M_{\Delta x}(x, t_{j_0})|$ differently starting from (8.27) as follows:

$$\begin{aligned}
|M_{\Delta x}(x, t_j) &- M_{\Delta x}(x, t_{j_0})| \\
&\leq \frac{\kappa F_2^*(G_2 \cdot t_j)}{\sqrt{2} K_0} \left[\int_{r_0}^{R} + \int_{R}^{\infty} \right] \rho_{\Delta x}(r + j\Delta x, 0) \|\Delta \mathbf{z}\| r^2 \, dr, \\
&\leq \frac{\kappa F_2^*(G_2 \cdot t_j) \bar{S} R^2}{\sqrt{2} K_0 L} G_2 |t_j - t_{j_0}| \quad (8.31) \\
&\quad + \frac{\kappa F_2^*(G_2 \cdot t_j) F_0^*(G_2 \cdot t_j)}{\sqrt{2} K_0} \int_{R}^{\infty} \rho_{\Delta x}(r + j\Delta x, 0) r^2 \, dr,
\end{aligned}$$

where we have used (8.28) together with

$$\|\Delta \mathbf{z}\| \leq F_0^*(G_2 \cdot t_j), \quad (8.32)$$

a consequence of (7.88). But

$$\begin{aligned}
\frac{\kappa}{2} \int_R^\infty \rho_{\Delta x}(r + j\Delta x, 0) r^2 \, dr &\leq \frac{\kappa}{2} \int_R^\infty \rho_{\Delta x}(r + j\Delta x, 0)(r + j\Delta x)^2 \, dr \\
&\leq \frac{\kappa}{2} \int_R^\infty \rho_{\Delta x}(r, 0) r^2 \, dr \\
&\leq \frac{\kappa}{2} \int_R^\infty u_{\Delta x}^0(r, 0) r^2 \, dr \\
&\leq M_{\Delta x}(\infty, 0) - M_{\Delta x}(R, 0), \quad (8.33)
\end{aligned}$$

and since

$$\lim_{R \to \infty} [M_{\Delta x}(R, 0) - M_{\Delta x}(\infty, 0)] = 0, \quad (8.34)$$

it follows that for any $\delta > 0$ sufficiently small, there exists $R(\delta) > 0$, such that

$$\frac{\kappa F_2^*(G_2 \cdot t_j) F_0^*(G_2 \cdot t_j)}{\sqrt{2} K_0} \int_{R(\delta)}^\infty \rho_{\Delta x}(r + j\Delta x, 0) r^2 \, dr \leq \delta, \quad (8.35)$$

for all $t_j \leq T_0$. Indeed, since $M_{\Delta x}(x, 0)$ is a continuous monotone increasing function of x, it follows that we can define $R(\delta)$ to satisfy the equality

$$\frac{\sqrt{2}}{K_0} F_2^*(G_2) F_0^*(G_2) [M_{\Delta x}(\infty, 0) - M_{\Delta x}(R(\delta), 0)] = \delta, \quad (8.36)$$

in which case (8.35) follows at once from (8.33). Using this definition of $R(\delta)$ in (8.31), it follows that for every $\delta > 0$,

$$|M_{\Delta x}(x, t_j) - M_{\Delta x}(x, t_{j_0})| \leq \frac{\kappa F_2^*(G_2 \cdot t_j) \bar{S} R(\delta)^2}{\sqrt{2} K_0 L} G_2 |t_j - t_{j_0}| + \delta. \quad (8.37)$$

Therefore, assuming Corollary 2, Theorem 4 applies up to some time T_0, $0 < T_0 \leq 1$, we can choose $\delta = \epsilon/2$, and set

$$T_\epsilon = \left\{ \frac{\kappa F_2^*(G_2) \bar{S} R(\epsilon/2)^2}{\sqrt{2} K_0 L} G_2 \right\}^{-1} \frac{\epsilon}{2}, \quad (8.38)$$

and conclude from (8.37) that

$$|M_{\Delta x}(x, t_j) - M_{\Delta x}(x, t_{j_0})| < \epsilon, \quad (8.39)$$

for all $t_j \leq Max\{T_\epsilon, T_0\}$. We record this as another corollary to Theorem 4:

Corollary 4 *Assume that Corollary 2, Theorem 4, holds up to time T_0. Then for all $\epsilon > 0$, there exists $T_\epsilon > 0$, (given explicitly in (8.38)), such that*

$$|M_{\Delta x}(x, t_j) - M_{\Delta x}(x, t_{j_0})| < \epsilon, \quad (8.40)$$

for all $x \geq r_0$, $t_j \leq Min\{T_\epsilon, T_0\}$.

We now use Corollary 4 to define $\bar{B}$ and T_B. Consider the function $B_{\Delta x}(x, t)$. Assume that the initial data satisfies

$$B_{\Delta x}(x, 0) = \frac{1}{1 - \frac{2 M_{\Delta x}(x,0)}{x}} \leq \bar{B}_0 \quad (8.41)$$

for some positive constant $\bar{B}_0$. Choose $\bar{B} > \bar{B}_0$, say

$$\bar{B} = 2\bar{B}_0. \quad (8.42)$$

Choose $\epsilon > 0$ by

$$\epsilon = Sup\left\{ \epsilon : \frac{1}{1 - \frac{2(M_{\Delta x}(x,0)+\epsilon)}{x}} \leq 2\bar{B}_0 = \bar{B}, \; for \; all \; r_0 \leq x < \infty \right\}. \quad (8.43)$$

Claim: By (8.43),

$$\epsilon \geq \frac{r_0}{2}\left(\frac{1}{\bar{B}_0} - \frac{1}{2\bar{B}_0}\right) > 0. \quad (8.44)$$

SHOCK-WAVE SOLUTIONS OF THE EINSTEIN EQUATIONS

To see this, let $\epsilon(x)$ be defined so that

$$\frac{1}{1 - \frac{2(M_{\Delta x}(x,0) + \epsilon(x))}{x}} = 2\bar{B}_0. \tag{8.45}$$

Solving (8.45) for $\epsilon(x)$ gives

$$\epsilon(x) = \frac{x}{2}\left\{1 - \frac{1}{2\bar{B}_0} - \frac{2M_{\Delta x}(x,0)}{x}\right\}. \tag{8.46}$$

But (8.41) implies

$$\frac{2M_{\Delta x}(x,0)}{x} \leq 1 - \frac{1}{\bar{B}_0}. \tag{8.47}$$

Using (8.47) in (8.46) gives (8.44). $\square$

Now for ϵ in (8.43), define

$$T_B = T_\epsilon \equiv \left\{\frac{\kappa F_2^*(G_2)\bar{S}R(\epsilon/2)^2}{\sqrt{2}K_0 L}G_2\right\}^{-1}\frac{\epsilon}{2}, \tag{8.48}$$

so that by (8.38), (8.40),

$$|M_{\Delta x}(x,t_j) - M_{\Delta x}(x,0)| < \epsilon, \tag{8.49}$$

for all $t_j \leq Max\{T_\epsilon, T_0\}$. But (8.40), (8.43), directly imply

$$\frac{1}{1 - \frac{2(M_{\Delta x}(x,0) + \epsilon)}{x}} \leq \bar{B}. \tag{8.50}$$

We conclude that assumption (7.30) of Corollary 2, Theorem 4, can be replaced by the condition that $\bar{B}$ is defined in (8.42), together with the condition that $t_j \leq T_B$, where T_B is defined in (8.48). We have shown that assumptions (7.29)-(7.32) of Corollary 2, Theorem 4, can be removed, and are consequences of appropriately restricting the time T_0 and redefining the constants involved in terms of the initial data.

The following theorem, which summarizes our results, follows directly from our construction of $\bar{v}, \bar{S}, \bar{M}, \bar{B}$ and T_S, T_M, T_B above:

Theorem 5 *Let $u_{\Delta x}(x,t)$, $\mathbf{A}_{\Delta x}(x,t)$ be an approximate solution generated by the fractional step Glimm method starting from initial data $u_{\Delta x}(x,0)$, $\mathbf{A}_{\Delta x}(x,0)$, and let $\bar{M}_0, \bar{B}_0, \bar{S}_0, \bar{v}_0$ and $\bar{V}_0$ be positive constants such that the initial data satisfies:*

$$M_{\Delta x}(x,0) \leq \bar{M}_0, \qquad (8.51)$$
$$B_{\Delta x}(x,0) \leq \bar{B}_0, \qquad (8.52)$$
$$0 < S_{\Delta x}(x,0) \leq \bar{S}_0 \qquad (8.53)$$
$$|v_{\Delta x}(x,0)| \leq \bar{v}_0 < c, \qquad (8.54)$$

for all $x \geq r_0$, and

$$\sum_{i_1 \leq i \leq i_2,\ p=1,2} |\gamma_{i0}^p| < V_0, \qquad (8.55)$$

for all $r_0 \leq x_{i_2} < x_{i_2} < \infty$, $|x_{i_2} - x_{i_1}| \leq L$. Let $\bar{v} = 2\bar{v}_0$, $\bar{S} = 2\bar{S}_0$, $\bar{M} = 2\bar{M}_0$, $\bar{B} = 2\bar{B}_0$, assume that

$$\frac{\Delta x}{\Delta t} = \Lambda = 2\sqrt{G_{AB}(\bar{B},\bar{M})}, \qquad (8.56)$$

and let

$$T = Min\left\{1, T_2, T_{\bar{S}}, T_{\bar{M}}, T_{\bar{B}}\right\}, \qquad (8.57)$$

where

$$\begin{aligned}
T_2 &= T_2 = \left(\frac{1}{G_2}\right) \frac{\bar{V}_*}{\left\{2\bar{V}_* + H(2\bar{V}_*)\right\}}, \qquad (8.58)\\
T_S &= Sup\{t : F_1^*(G_2 \cdot t_j) \leq 2F_1^*(0),\ \text{for all } t_j \leq t\},\\
T_M &= T_M = Sup\{t : M_{r_0} + F_2^*(G_2 \cdot t_j) M_0 \leq \bar{M},\ \text{for all } t_j \leq t\},\\
T_B &= \left\{\frac{\kappa F_2^*(G_2)\bar{S}R(\epsilon/2)^2}{\sqrt{2}K_0 L} G_2\right\}^{-1} \frac{\epsilon}{2},
\end{aligned}$$

and

$$\epsilon = Sup\left\{\epsilon : \frac{1}{1 - \frac{2(M_{\Delta x}(x,0)+\epsilon)}{x}} \leq \bar{B},\ \text{for all } r_0 \leq x < \infty\right\}, \qquad (8.59)$$

c.f., (7.84), (8.10), (8.18), (8.48) and (8.43). Then the approximate solution $u_{\Delta x}$, $\mathbf{A}_{\Delta x}$ is well defined for all $r_0 \leq r < \infty$, $0 \leq t \leq T$, and satisfies the bounds

$$M_{\Delta x}(x, t_j) \leq \bar{M}, \tag{8.60}$$
$$B_{\Delta x}(x, t_j) \leq \bar{B}, \tag{8.61}$$
$$0 < S_{\Delta x}(x, t_j) \leq \bar{S} \tag{8.62}$$
$$|v_{\Delta x}(x, t_j)| \leq \bar{v} < c, \tag{8.63}$$

together with the bounds

$$\sum_{i_1 \leq i \leq i_2,\ p=1,2} |\gamma_{ij}^p| < 2\bar{V}_* = 2\left(1 + \frac{4\sqrt{G_{AB}}}{L}\right) V_0, \tag{8.64}$$

$$\|\mathbf{z}_{ij} - \mathbf{z}_{i+j,0}\| \leq F_0^*(G_2 \cdot T), \tag{8.65}$$

$$\int_{x_{i_1}}^{x_{i_2}} \|\mathbf{z}_{\Delta x}(x, t_{j_2}) - \mathbf{z}_{\Delta x}(x, t_{j_1})\|\, dx \leq G_2 |t_{j_2} - t_{j_1}|, \tag{8.66}$$

$$|A'_{\Delta x}(x, t_j)| \leq \left(\frac{1}{r_0} + \kappa \frac{c^2 + \sigma^2 \bar{v}^2}{c^2 - \bar{v}^2} \bar{S}\right) \bar{B} G_{AB}, \tag{8.67}$$

$$|B'_{\Delta x}(x, t_j)| \leq \left(\frac{1}{r_0} + \kappa \frac{c^2 + \sigma^2 \bar{v}^2}{c^2 - \bar{v}^2} \bar{S}\right) \bar{B}^2, \tag{8.68}$$

$$|M_{\Delta x}(x, t_{j_2}) - M_{\Delta x}(x, t_{j_1})| \leq \frac{\kappa F_2^*(G_2 \cdot T) \bar{S} x^2}{\sqrt{2} K_0 L} G_2 |t_{j_2} - t_{j_1}|, \tag{8.69}$$

for all $r_0 \leq x, x_{i_1}, x_{i_2} < \infty$, $|x_{i_2} - x_{i_1}| \leq L$, and $0 \leq t_j, t_{j_1}, t_{j_2} \leq T$, c.f. (7.83), (7.90), (7.41), (8.30), (7.86), (7.43), (7.44).

Recall that the constants $G_{AB} \equiv G_{AB}(\bar{M}, \bar{B})$, $G_2 \equiv G_2(\bar{M}, \bar{B}, \bar{S})$, $F_1^*(G_2 \cdot T)$, $F_2^*(G_2 \cdot T)$, and $V_*(G_2 \cdot T)$, defined in (3.43),(7.81),(8.7),(8.15), and (7.85) respectively, are based on the functions $G_{AB}(\cdot, \cdot)$, $G_2(\cdot, \cdot, \cdot)$, $F_i^*(\cdot)$ and $V_*(\cdot)$ that depend only on the constants σ, c, K_0, r_0, L and V_0, and thus are determined by the initial data alone.

Corollary 5 *Let $u_{\Delta x}(x, t)$, $\mathbf{A}_{\Delta x}(x, t)$ be an approximate solution generated by the fractional step Glimm method starting from initial data $u_{\Delta x}(x, 0)$, $\mathbf{A}_{\Delta x}(x, 0)$, that satisfies the conditions (8.60)-(8.59) of Theorem 5. Then there exists a subsequence $\Delta x \to 0$ and bounded measurable functions $u(x, t) = \Psi^{-1} \cdot \Phi \cdot \mathbf{z}(x, t)$, $\mathbf{A}(x, t)$, such that $(u_{\Delta x}, \mathbf{A}_{\Delta x}) \to (u, \mathbf{A})$ for a.e. $(x, t) \in [r_0, \infty) \times [0, T]$. Moreover, the convergence $u_{\Delta x}(\cdot, t) \to u(\cdot, t)$ is in L^1_{loc} for* each $t \in [0, T]$, *uniformly on compact sets in (x, t)-space, and the limit function $u_{\Delta x}$ satisfies:*

$$\begin{aligned}
TV_{[x_1, x_2]} z(\cdot, t) &\leq 2\bar{V}_*, \\
TV_{[x_1, x_2]} w(\cdot, t) &\leq H(2\bar{V}_*), \\
TV_{[x_1, x_2]} \mathbf{z}(\cdot, t) &\leq 2\bar{V}_* + H(2\bar{V}_*),
\end{aligned} \quad (8.70)$$

$$\|\mathbf{z}(x, t) - \mathbf{z}(x + \lambda T, 0)\| \leq F_0^*(G_2 \cdot T), \quad (8.71)$$

and

$$\int_{x_1}^{x_2} \|\mathbf{z}(x, t_2) - \mathbf{z}(x, t_1)\| \, dx \leq G_2 |t_2 - t_1|, \quad (8.72)$$

for all $r_0 \leq x, x_1, x_2 < \infty$, $|x_2 - x_1| < L$, and $0 \leq t, t_1, t_2 \leq T$.

The convergence in $\mathbf{A}$ is pointwise a.e., uniformly on compact sets in (x, t)-space, and the limit function $\mathbf{A}(x, t)$ satisfies

$$A(x, t) = A_{r_0} \exp \int_{r_0}^{x} \left\{ \frac{B(r, t) - 1}{r} + \kappa r B(r, t) T_M^{11}(u(r, t)) \right\} dr, \quad (8.73)$$

$$B(r, t) = \frac{1}{1 - \frac{2M(r,t)}{r}}, \quad M(r, t) = M(r_0, t) + \frac{\kappa}{2} \int_{r_0}^{r} u^0(r, t) r^2 \, dr, \quad (8.74)$$

$$\left| \frac{A(x + y, t) - A(x, t)}{y} \right| \leq \left(\frac{1}{r_0} + \kappa \frac{c^2 + \sigma^2 \bar{v}^2}{c^2 - \bar{v}^2} \bar{S} \right) \bar{B} G_{AB}, \quad (8.75)$$

$$\left| \frac{B(x + y, t) - B(x, t)}{y} \right| \leq \left(\frac{1}{r_0} + \kappa \frac{c^2 + \sigma^2 \bar{v}^2}{c^2 - \bar{v}^2} \bar{S} \right) \bar{B}^2, \quad (8.76)$$

$$|M(x, t_2) - M(x, t_1)| \leq \frac{\kappa F_2^*(G_2 \cdot T) \bar{S} x^2}{\sqrt{2} K_0 L} G_2 |t_2 - t_1|, \quad (8.77)$$

for all $r_0 \leq x, x_1, x_2 < \infty$, $|x_2 - x_1| \leq L$, and $0 \leq t, t_1, t_2 \leq T$.

Proof: It follows from (7.90), (8.64) (together with the non-singularity of the mapping from $\mathbf{z} \to u$) that the approximate solution $u_{\Delta x}(x, t)$ is bounded, and of locally bounded total variation at each fixed time $0 \leq t \leq T$, and these bounds are uniform in time over compact x-intervals. Moreover, it follows from (8.66) that $u_{\Delta x}(x,t)$ is locally Lipschitz continuous in the L^1-norm at each time, uniformly on compact sets. These bounds are uniform as $\Delta x \to 0$. This is all that is required to apply Oleinik's compactness argument to the function $u_{\Delta x}$, [8, 23, 16]. From this we can conclude that there exists a sequence $\Delta x \to 0$ such that $u_{\Delta x}$ converges a.e. to a bounded measurable function u on $x \geq r_0, 0 \leq t \leq T$. The convergence is in L^1_{loc} at each time, uniformly on compact sets, and the supnorm bound (8.71), the local total variation estimate (8.70), and the continuity of the local L^1 norm (8.72), carry over from the corresponding estimates (8.71), (8.64), (7.86) for approximate solution. (For (8.64) we use that the change Δw across a wave is bounded by $H(\Delta z)$, and that H is a convex function, c.f. Proposition 5. The Oleinik argument is based on using Helly's Theorem to extract a pointwise convergent subsequence on a dense set of times between $t = 0$ and $T = t$, and then to use the local L^1-Lipschitz continuity of $u_{\Delta x}$ to extrapolate the L^1 convergence to all intermediate times, [16].)

It follows from (8.69)-(8.68), together with (3.28), that $\mathbf{A}_{\Delta x}$ is locally Lispchitz continuous in x and t for $x \geq r_0$, $t \leq T$, (ignoring errors that are of order Δx), and the Lipschitz bounds are uniform as $\Delta x \to 0$. It follows from Arzela-Ascoli that on some subsequence $\Delta x \to 0$, $\mathbf{A}_{\Delta x}$ converges to a locally Lipschitz continuous function $\mathbf{A}(x,t)$, and the convergence is pointwise almost everywhere, uniformly on compact sets. It follows that the convergence of $u_{\Delta x}$ and $\mathbf{A}_{\Delta x}$ is strong enough to pass the limit through the integral sign in (3.28) and (1.18), and thus conclude (8.73) and (8.74), respectively. Similarly, (8.76)-(8.77) are obtained from (7.43)-(8.30), respectively. The initial data u_0 is taken on in the L^1 sense,

$$\lim_{t \to 0} \|u(\cdot, t) - u_0(\cdot)\|_{L^1_{loc}} = 0, \tag{8.78}$$

and the boundary condition $v = 0 \iff M(r_0, 0) = M_{r_0}$ is taken on weakly, c.f. [16]. $\square$

Proof of Theorem 3: In the final section we prove that for almost every sample sequence $\mathbf{a}$, the functions $u_{\Delta x}(x,t), \mathbf{A}(x,t)$ define a weak solution of the Einstein equations (1.2)-(1.5) on $r_0 \leq x < \infty, 0 \leq t \leq T$. Assuming this, $u_{\Delta x}(x,t), \mathbf{A}(x,t)$ is then a weak solution of (1.2)-(1.5) in the class $u_{\Delta x}$ bounded measurable and $\mathbf{A}_{\Delta x}$ Lipschitz continuous, and so it follows that our results in [10] apply. In particular, (1.3) holds in the pointwise almost everywhere sense. Thus the proof of Theorem 3 is complete once we verify (2.42). (The assumptions (2.28)-(2.29) just imply that $TV_{[x_1,x_2]}\mathbf{z}(\cdot, 0) < \infty$, and this guarantees (8.64).) For (2.42), note first that (1.24) together with (8.71) imply that

$$\lim_{x \to 0} \dot{M}(x,t) = 0, \tag{8.79}$$

for all $0 \leq t \leq T$. To see this, recall from Theorem 2 that if $u_{\Delta x}, \mathbf{A}_{\Delta x}$ is a weak solution for $0 \leq t \leq T$, then (1.3) and (1.8) hold. By (1.8), statement (8.79) follows so long as

$$\lim_{r \to \infty} \sqrt{\frac{A(r,t)}{B(r,t)}} u^1(r,t) r^2 = 0 \tag{8.80}$$

for $t \leq T$, where

$$\left| \sqrt{\frac{A(r,t)}{B(r,t)}} u^1(r,t) r^2 \right| \leq \sqrt{A(r,t)} u^0(r,t) r^2. \tag{8.81}$$

Now since A and B are given by (8.73) and (8.74), it follows that $\mathbf{A}$ satisfies (1.2) and (1.4), and so adding these two equations, and following the argument leading to (3.43), we obtain that

$$|A| \leq A_{r_0} B_{r_0} exp\left\{\frac{8\bar{B}\bar{M}}{r_0}\right\},$$

and thus A is uniformly bounded. Since $|v(x,t)| \leq \bar{v} < c$, (1.24) and (8.71) imply that

$$\lim_{r \to \infty} \sqrt{A(r,t)} u^0(r,t) r^2 = 0, \tag{8.82}$$

and so (8.79) follows as claimed. But (8.79) implies that,

$$\lim_{x \to \infty} M(x,t) = \lim_{x \to \infty} M(x,0) = M_\infty, \tag{8.83}$$

which is (2.42) of Theorem 3. We conclude from Theorem 2 that the proof of Theorem 3 is complete once we prove that $u(x,t), \mathbf{A}(x,t)$ is a genuine weak solution of (1.26),(1.27) with initial boundary data 2.36)-(2.38). This is the topic of the next section. $\square$

Our theorems have the following corollary:

Corollary 6 *Assume that the initial data $u_0(x)$ satisfies (1)-(5). Then a bounded weak solution $u(x,t), \mathbf{A}(x,t)$ of the Einstein equations (1.2)-(1.5) exists up until the first time T at which either*

$$\lim_{t \to T^-} Sup_x B(x,t) = \infty, \tag{8.84}$$

$$\lim_{t \to T^-} Sup_x x\rho(x,t) = \infty, \tag{8.85}$$

or

$$\lim_{t \to T^-} Sup_x TV_{[x_1,x_2]}\mathbf{z}(\cdot,t) = \infty. \tag{8.86}$$

Proof: If B, S and $TV_{[x_1,x_2]}\mathbf{z}$ remain uniformly bounded up to time T, then our argument shows that v remains uniformly bounded away from c up to time T, c.f. (8.2)-(8.3). Thus we can repeat the proof that the solution starting from initial data at time T, continues forward for some positive time. The Corollary follows at once.

9 Convergence

In this section we prove that the approximate solutions $u_{\Delta x}, \mathbf{A}_{\Delta x}$ which satisfy the estimates (8.70)-(8.77) of Corollary 5, Theorem 5, are weak solutions of (1.30), (1.31), for almost every choice of sample sequence $\mathbf{a} \in \Pi$, c.f. (3.24). This is a modification of Glimm's original argument [8], and the argument in [16]. The main point is to show that the the discontinuities in $\mathbf{A}_{\Delta x}$ at the boundary of the mesh rectangles $\mathcal{R}_{ij}$ are accounted for by inclusion of the term

$$\mathbf{A}' \cdot \nabla_{\mathbf{A}} f(\mathbf{A}, u, x) = \frac{1}{2}\sqrt{\frac{A}{B}} \delta\left(T_M^{01}, T_M^{11}\right),$$

in the ODE step (3.22), c.f. (3.17).

To start, recall that $u_{\Delta x}^{RP}$ denotes the exact Riemann problem solution in each $\mathcal{R}_{ij}$ for the homogeneous system (4.1), so that

$$\begin{aligned} 0 = & \iint_{\mathcal{R}_{ij}} \left\{ -u_{\Delta x}^{RP} \varphi_t - f(\mathbf{A}_{ij}, u_{\Delta x}^{RP})\varphi_x \right\} dx dt \\ & + \int_{\mathcal{R}_i} \left\{ u_{\Delta x}^{RP}(x, t_{j+1}^-)\varphi(x, t_{j+1}) - u_{\Delta x}^{RP}(x, t_j^+)\varphi(x, t_j) \right\} dx \quad (9.1) \\ & + \int_{\mathcal{R}_j} \left\{ f(\mathbf{A}_{ij}, u_{\Delta x}^{RP}(x_{i+\frac{1}{2}}, t))\varphi(x_{i+\frac{1}{2}}, t) \right. \\ & \left. - f(\mathbf{A}_{ij}, u_{\Delta x}^{RP}(x_{i-\frac{1}{2}}, t))\varphi(x_{i-\frac{1}{2}}, t) \right\} dt. \end{aligned}$$

Recall that $\hat{u}(t, u_0)$ denotes the solution to the initial value problem

$$\hat{u}_t = G(\mathbf{A}_{ij}, \hat{u}, x) = g(\mathbf{A}_{ij}, \hat{u}, x) - \mathbf{A}' \cdot \nabla_{\mathbf{A}} f(\mathbf{A}_{ij}, \hat{u}, x),$$
$$\hat{u}(0) = u_0.$$

Thus

$$\begin{aligned}\hat{u}(t, u_0) - u_0 &= \int_0^t \hat{u}_t \, dt \\ &= \int_0^t \{g(\mathbf{A}_{ij}, \hat{u}(\xi, u_0), x) - \mathbf{A}' \cdot \nabla_{\mathbf{A}} f(\mathbf{A}_{ij}, \hat{u}, x)\} \, dt.\end{aligned}$$

Since $\hat{u}$ implements the ODE step of the fractional step method, it follows that the approximate solution $u_{\Delta x}(x, t)$ is defined on each mesh rectangle $\mathcal{R}_{ij}$ by the formula

$$\begin{aligned}u_{\Delta x}(x, t) &= u_{\Delta x}^{RP}(x, t) + \int_{t_j}^t \{g(\mathbf{A}_{ij}, \hat{u}(\xi - t_j, u_{\Delta x}^{RP}(x,t)), x)\} \, dt \qquad (9.2)\\ &\quad - \int_{t_j}^t \left\{\frac{\partial f}{\partial \mathbf{A}}(\mathbf{A}_{ij}, \hat{u}(\xi - t_j, u_{\Delta x}^{RP}(x,t))) \cdot \mathbf{A}'_{\Delta x}\right\} dt.\end{aligned}$$

Note that the difference between the approximate and Riemann problem solutions is on the order of Δx. Define the residual $\epsilon(u_{\Delta x}, \mathbf{A}_{\Delta x}, \varphi)$ of the approximate solutions $u_{\Delta x}$ by

$$\begin{aligned}\epsilon_{\Delta x} &\equiv \epsilon(u_{\Delta x}, \mathbf{A}_{\Delta x}, \varphi) \\ &= \int_{r_0}^\infty \int_0^\infty \{-u_{\Delta x}\varphi_t - f(\mathbf{A}_{\Delta x}, u_{\Delta x})\varphi_x - g(\mathbf{A}_{\Delta x}, u_{\Delta x}, x)\varphi\} \, dtdx \\ &\quad - I_1 - I_2, \\ &= \sum_{ij} \int\int_{\mathcal{R}_{ij}} \{-u_{\Delta x}\varphi_t - f(\mathbf{A}_{ij}, u_{\Delta x})\varphi_x - g(\mathbf{A}_{\Delta x}, u_{\Delta x}, x)\varphi\} \, dtdx \\ &\quad - I_1 - I_2, \qquad (9.3)\end{aligned}$$

where

$$I_1 = \int_{r_0}^\infty u_{\Delta x}(x, 0^+)\varphi(x, 0) \, dx = \sum_i \int_{\mathcal{R}_i} u_{\Delta x}(x, 0^+)\varphi(x, 0) \, dx, \qquad (9.4)$$

and

$$I_2 = \int_0^\infty f(\mathbf{A}_{\Delta x}(r_0^+,t), u_{\Delta x}(r_0^+,t))\varphi(r_0,t)\,dt$$
$$= \sum_j \int_{\mathcal{R}_j} f(\mathbf{A}_{ij}, u_{\Delta x}(r_0^+,t))\varphi(r_0,t)\,dt. \tag{9.5}$$

We now prove that the residual is $O(\Delta x)$. (It follows that if $u_{\Delta x} \to u$ and $\mathbf{A}_{\Delta x} \to \mathbf{A}$ converge in L^1_{loc} at each time, uniformly on compact sets, then the limit function will satisfy $\epsilon(u, \mathbf{A}, \varphi) = 0$, the condition that u be a weak solution of the Einstein equations.) To this end, substitute (9.2) into (9.3) to obtain

$$\begin{aligned}\epsilon(u_{\Delta x}, \mathbf{A}_{\Delta x}, \varphi) &= \sum_{ij} \int\!\!\int_{\mathcal{R}_{ij}} \Big\{ -u^{RP}_{\Delta x}\varphi_t - f(\mathbf{A}_{ij}, u_{\Delta x})\varphi_x - g(\mathbf{A}_{ij}, u_{\Delta x}, x)\varphi \\ &\quad -\varphi_t \int_{t_j}^t \Big[g(\mathbf{A}_{ij}, \hat{u}(\xi - t_j, u^{RP}_{\Delta x}(x,t)), x) \\ &\quad - \frac{\partial f}{\partial \mathbf{A}}(\mathbf{A}_{ij}, \hat{u}(\xi - t_j, u^{RP}_{\Delta x}(x,t))) \cdot \mathbf{A}'_{\Delta x} \Big] d\xi \Big\}\, dx\,dt \\ &\quad -I_1 - I_2.\end{aligned} \tag{9.6}$$

Set

$$\begin{aligned}I^1_{ij}(x,t) &= \int_{t_j}^t \Big[g(\mathbf{A}_{ij}, \hat{u}(\xi - t_j, u^{RP}_{\Delta x}(x,t)), x) \Big] d\xi \\ &\quad \int_{t_j}^t \Big[-\frac{\partial f}{\partial \mathbf{A}}(\mathbf{A}_{ij}, \hat{u}(\xi - t_j, u^{RP}_{\Delta x}(x,t))) \cdot \mathbf{A}'_{\Delta x} \Big] d\xi.\end{aligned}$$

Upon substituting (9.1) into (9.7), we have

$$\begin{aligned}\epsilon_{\Delta x} &= \sum_{ij}\int\!\!\int_{\mathcal{R}_{ij}} \Big\{ \varphi_x \big[f(\mathbf{A}_{ij}, u^{RP}_{\Delta x}) - f(\mathbf{A}_{ij}, u_{\Delta x}) \big] - g(\mathbf{A}_{ij}, u_{\Delta x}, x)\varphi \\ &\quad - \varphi_t I^1_{ij}(x,t) \Big\} dx\,dt \\ &\quad -I_1 - \sum_{ij}\int_{\mathcal{R}_i} \Big\{ u^{RP}_{\Delta x}(x, t_{j+1}^-)\varphi(x, t_{j+1}) - u^{RP}_{\Delta x}(x, t_j^+)\varphi(x, t_j) \Big\} dx \\ &\quad -I_2 - \sum_{ij}\int_{\mathcal{R}_j} \Big\{ f(\mathbf{A}_{ij}, u^{RP}_{\Delta x}(x_{i+\frac{1}{2}}, t))\varphi(x_{i+\frac{1}{2}}, t) \\ &\quad - f(\mathbf{A}_{ij}, u^{RP}_{\Delta x}(x_{i-\frac{1}{2}}, t))\varphi(x_{i-\frac{1}{2}}, t) \Big\} dt\end{aligned} \tag{9.7}$$

Note that

$$|f(\mathbf{A}_{ij}, u_{\Delta x}^{RP}) - f(\mathbf{A}_{ij}, u_{\Delta x})| \leq C\Delta t, \qquad (9.8)$$

and so

$$|\sum_{ij} \int\int_{\mathcal{R}_{ij}} \varphi \left[f(\mathbf{A}_{ij}, u_{\Delta x}^{RP}) - f(\mathbf{A}_{ij}, u_{\Delta x})\right] dx dt| \leq |\varphi_x|_\infty C \Delta t T(b-a), \qquad (9.9)$$

where $Supp(\varphi) \subset [a,b] \times [0.T]$. (We let C denote a generic constant that depends only on the bounds for the solution.) Using the fact that $u_{\Delta x}^{RP}(x, t_j^+) = u_{\Delta x}(x, t_j^+)$, and inserting (9.2), we obtain that

$$-I_1 - \sum_{ij} \int_{\mathcal{R}_i} \left\{ u_{\Delta x}^{RP}(x, t_{j+1}^-) \varphi(x, t_{j+1}) - u_{\Delta x}^{RP}(x, t_j^+) \varphi(x, t_j) \right\} dx$$

$$= \sum_{j \neq 0} \int_{r_0}^{\infty} \left\{ u_{\Delta x}(x, t_j^+) - u_{\Delta x}^{RP}(x, t_j^-) \right\} \varphi(x, t_j) dx$$

$$= \sum_{j \neq 0} \int_{r_0}^{\infty} \varphi(x, t_j) \left\{ u_{\Delta x}(x, t_j^+) - u_{\Delta x}(x, t_j^-) \right\} dx \qquad (9.10)$$

$$+ \sum_{j \neq 0} \int_{r_0}^{\infty} \varphi(x, t_j) \left\{ u_{\Delta x}(x, t_j^-) - u_{\Delta x}^{RP}(x, t_j^-) \right\} dx.$$

Set

$$\epsilon_1(u_{\Delta x}, \mathbf{A}_{\Delta x}, \varphi) = \sum_{j \neq 0} \int_{r_0}^{\infty} \varphi(x, t_j) \left\{ u_{\Delta x}(x, t_j^+) - u_{\Delta x}(x, t_j^-) \right\} dx. \qquad (9.11)$$

It follows that

$$\begin{aligned}\epsilon_{\Delta x} &= \epsilon_1(u_{\Delta x}, \mathbf{A}_{\Delta x}, \varphi) + \sum_{ij} \int\int_{\mathcal{R}_{ij}} \left\{ -g(\mathbf{A}_{ij}, u_{\Delta x}, x)\varphi - \varphi_t I_{ij}^1(x, t) \right\} dx dt \\ &+ \sum_{j \neq 0} \int_{r_0}^{\infty} \varphi(x, t_j) \left\{ u_{\Delta x}(x, t_j^-) - u_{\Delta x}^{RP}(x, t_j^-) \right\} dx \qquad (9.12) \\ &- I_2 - \sum_{ij} \int_{\mathcal{R}_j} \left\{ f(\mathbf{A}_{ij}, u_{\Delta x}^{RP}(x_{i+\frac{1}{2}}, t))\varphi(x_{i+\frac{1}{2}}, t) \right. \\ &\left. - f(\mathbf{A}_{ij}, u_{\Delta x}^{RP}(x_{i-\frac{1}{2}}, t))\varphi(x_{i-\frac{1}{2}}, t) \right\} dt + O(\Delta x).\end{aligned}$$

But adding $-I_2$ to

$$-\sum_{ij} \int_{\mathcal{R}_j} \left\{ f(\mathbf{A}_{ij}, u_{\Delta x}^{RP}(x_{i+\frac{1}{2}}, t))\varphi(x_{i+\frac{1}{2}}, t) - f(\mathbf{A}_{ij}, u_{\Delta x}^{RP}(x_{i-\frac{1}{2}}, t))\varphi(x_{i-\frac{1}{2}}, t) \right\} dt$$

gives

$$\sum_{ij} \int_{\mathcal{R}_j} \left\{ f(\mathbf{A}_{i+1,j}, u_{\Delta x}^{RP}(x_{i+\frac{1}{2}}, t)) - f(\mathbf{A}_{ij}, u_{\Delta x}^{RP}(x_{i+\frac{1}{2}}, t)) \right\} \varphi(x_{i+\frac{1}{2}}, t) \, dt \quad (9.13)$$

$$+ \sum_{j} \int_{\mathcal{R}_j} \left\{ f(\mathbf{A}_{0j}, u_{\Delta x}^{RP}(r_0^+, t)) - f(\mathbf{A}_{0j}, u_{\Delta x}(r_0^+, t)) \right\} \varphi(r_0, t) \, dt,$$

where

$$\left| \sum_{j} \int_{\mathcal{R}_j} \left\{ f(\mathbf{A}_{0j}, u_{\Delta x}^{RP}(r_0^+, t)) - f(\mathbf{A}_{0j}, u_{\Delta x}(r_0^+, t)) \right\} \varphi(r_0, t) \, dt \right|$$

$$\leq |\varphi|_\infty C \Delta t^2 \left(\frac{T}{\Delta t} \right) = O(\Delta t). \quad (9.14)$$

To analyze the term multiplied by φ_t in (9.12), we add and subtract a term that differs from this one by $O(\Delta x)$, and then use integration by parts on the new term. That is, set $I_{\Delta S}$ equal to the expression

$$\sum_{ij} \int \int_{\mathcal{R}_{ij}} \varphi_t \int_{t_j}^{t} \Big[g(\mathbf{A}_{ij}, \hat{u}(\xi - t_j, u_{\Delta x}^{RP}(x, \xi)), x)$$

$$- g(\mathbf{A}_{ij}, \hat{u}(\xi - t_j, u_{\Delta x}^{RP}(x, t)), x) - \frac{\partial f}{\partial \mathbf{A}}(\mathbf{A}_{ij}, \hat{u}(\xi - t_j, u_{\Delta x}^{RP}(x, \xi))) \cdot \mathbf{A}'_{\Delta x}$$

$$+ \frac{\partial f}{\partial \mathbf{A}}(\mathbf{A}_{ij}, \hat{u}(\xi - t_j, u_{\Delta x}^{RP}(x, t))) \cdot \mathbf{A}'_{\Delta x} \Big] d\xi \, dx dt.$$

But

$$|I_{\Delta S}| \leq \sum_{ij} \int \int_{\mathcal{R}_{ij}} |\varphi_t|_\infty \int_{t_j}^{t} C |\gamma_{ij}^l| \, d\xi \, dx dt|$$

$$\leq |\varphi_t|_\infty C \Delta t^2 \Delta x \sum_{ijl} |\gamma_{ij}^l|$$

$$\leq CV |\varphi_t|_\infty \Delta t^2 \Delta x \frac{T}{\Delta t} = O(\Delta x^2),$$

and so

$$- \sum_{ij} \int \int_{\mathcal{R}_{ij}} \varphi_t \int_{t_j}^{t} \Big[g(\mathbf{A}_{ij}, \hat{u}(\xi - t_j, u_{\Delta x}^{RP}(x, t)), x)$$

$$- \frac{\partial f}{\partial \mathbf{A}}(\mathbf{A}_{ij}, \hat{u}(\xi - t_j, u_{\Delta x}^{RP}(x, t))) \cdot \mathbf{A}'_{\Delta x} \Big] d\xi . \, dx dt$$

$$
\begin{aligned}
&= I_{\Delta S} - \sum_{ij} \int\int_{\mathcal{R}_{ij}} \varphi_t \int_{t_j}^{t} \Big[g(\mathbf{A}_{ij}, \hat{u}(\xi - t_j, u^{RP}_{\Delta x}(x, \xi)), x) \\
&\qquad - \frac{\partial f}{\partial \mathbf{A}}(\mathbf{A}_{ij}, \hat{u}(\xi - t_j, u^{RP}_{\Delta x}(x, \xi))) \cdot \mathbf{A}'_{\Delta x} \Big] d\xi.\, dx dt \\
&= -\sum_{ij} \int_{\mathcal{R}_i} \Bigg\{ \varphi(x, t_{j+1}) \int_{t_j}^{t_{j+1}} \Big[g(\mathbf{A}_{ij}, \hat{u}(\xi - t_j, u^{RP}_{\Delta x}(x, \xi)), x) \\
&\qquad - \frac{\partial f}{\partial \mathbf{A}}(\mathbf{A}_{ij}, \hat{u}(\xi - t_j, u^{RP}_{\Delta x}(x, \xi))) \cdot \mathbf{A}'_{\Delta x} \Big] d\xi \\
&\qquad - \int_{t_j}^{t_{j+1}} \varphi \Big[g(\mathbf{A}_{ij}, u_{\Delta x}, x) - \frac{\partial f}{\partial \mathbf{A}}(\mathbf{A}_{ij}, u_{\Delta x}) \cdot \mathbf{A}'_{\Delta x} \Big] dt \Bigg\} dx \\
&\quad + O(\Delta x^2) \\
&= -\sum_{ij} \int_{\mathcal{R}_i} \Bigg\{ \varphi(x, t_{j+1}) \int_{t_j}^{t_{j+1}} \Big[g(\mathbf{A}_{ij}, \hat{u}(\xi - t_j, u^{RP}_{\Delta x}(x, t_{j+1})), x) \\
&\qquad - \frac{\partial f}{\partial \mathbf{A}}(\mathbf{A}_{ij}, \hat{u}(\xi - t_j, u^{RP}_{\Delta x}(x, t_{j+1}))) \cdot \mathbf{A}'_{\Delta x} \Big] d\xi \Bigg\} dx \qquad (9.15) \\
&\quad s + I_4 + I_5 + O(\Delta x^2)
\end{aligned}
$$

where

$$
\begin{aligned}
I_4 &= \sum_{ij} \int_{\mathcal{R}_i} \Bigg\{ \varphi(x, t_{j+1}) \int_{t_j}^{t_{j+1}} \Big[g(\mathbf{A}_{ij}, \hat{u}(\xi - t_j, u^{RP}_{\Delta x}(x, t_{j+1})), x) \\
&\qquad - g(\mathbf{A}_{ij}, \hat{u}(\xi - t_j, u^{RP}_{\Delta x}(x, \xi)), x) \\
&\qquad - \frac{\partial f}{\partial \mathbf{A}}(\mathbf{A}_{ij}, \hat{u}(\xi - t_j, u^{RP}_{\Delta x}(x, t_{j+1}))) \cdot \mathbf{A}'_{\Delta x} \\
&\qquad + \frac{\partial f}{\partial \mathbf{A}}(\mathbf{A}_{ij}, \hat{u}(\xi - t_j, u^{RP}_{\Delta x}(x, \xi))) \cdot \mathbf{A}'_{\Delta x} \Big] d\xi \Bigg\} dx,
\end{aligned}
$$

and

$$
I_5 = \sum_{ij} \int\int_{\mathcal{R}_{ij}} \varphi \Big[g(\mathbf{A}_{ij}, u_{\Delta x}, x) - \frac{\partial f}{\partial \mathbf{A}}(\mathbf{A}_{ij}, u_{\Delta x}) \cdot \mathbf{A}'_{\Delta x} \Big] dx dt. \qquad (9.16)
$$

Note that

$$
\begin{aligned}
|I_4| &\leq |\varphi|_\infty \sum_{ijl} C |\gamma^l_{ij}| \Delta x \Delta t \leq |\varphi|_\infty C \Delta x \Delta t \sum_j V \\
&= |\varphi|_\infty C \Delta x \Delta t V \frac{T}{\Delta t} = O(\Delta x),
\end{aligned}
$$

where the sum on j is taken over t_j in $Supp(\varphi)$. Substituting (9.13) and (9.15) into (9.12), we have

$$\epsilon(u_{\Delta x}, \mathbf{A}_{\Delta x}, \varphi) = O(\Delta x) + \epsilon_1(u_{\Delta x}, \mathbf{A}_{\Delta x}, \varphi)$$
$$- \sum_{ij} \int\int_{\mathcal{R}_{ij}} \varphi \frac{\partial f}{\partial \mathbf{A}}(\mathbf{A}_{ij}, u_{\Delta x}) \cdot \mathbf{A}'_{\Delta x} \, dxdt \qquad (9.17)$$
$$+ \sum_{ij} \int_{\mathcal{R}_j} \varphi(x_{i+\frac{1}{2}}, t) \left\{ f(\mathbf{A}_{i+1,j}, u_{\Delta x}^{RP}(x_{i+\frac{1}{2}}, t)) - f(\mathbf{A}_{ij}, u_{\Delta x}^{RP}(x_{i+\frac{1}{2}}, t)) \right\} \, dt.$$

It is evident now that

$$\epsilon(u_{\Delta x}, \mathbf{A}_{\Delta x}, \varphi) = \epsilon_1(u_{\Delta x}, \mathbf{A}_{\Delta x}, \varphi) + O(\Delta x). \qquad (9.18)$$

We use Glimm's technique to show that $\epsilon_1(u_{\Delta x}, \mathbf{A}_{\Delta x}, \varphi) = O(\Delta x)$, c.f. [8].

To estimate ϵ_1, write $\epsilon \equiv \epsilon_1(\Delta x, \varphi, \mathbf{a})$ to display its dependence on $(\Delta x, \phi, \mathbf{a})$, where $\mathbf{a} \in \Pi$ is the sample sequence, c.f. (3.24) above. Set

$$\epsilon_1^j(\Delta x, \varphi, \mathbf{a}) = \int_{r_0}^{\infty} \varphi(x, t_j) \left\{ u_{\Delta x}(x, t_j^+) - u_{\Delta x}(x, t_j^-) \right\} dx. \qquad (9.19)$$

Now since $u(x,t) = \Psi^{-1} \cdot \Phi \cdot \mathbf{z}(x,t)$, it follows from (8.70) that there exists a constant V such that $TV_{[x,x+L]}u_{\Delta x}(\cdot, t) \leq V$ on $r_0 \leq x < \infty$, $t < T$. Using this, the following lemma gives estimates for ϵ_1 and ϵ_1^j.

Lemma 5 *Let $\mathbf{a} \in \Pi$, and let $\varphi \in C_0 \cap L^{\infty}$ be a test function in the space of continuous functions of compact support in $r_0 \leq x < \infty$, $0 \leq t < T$. Suppose $TV_{[x,x+L]}u_{\Delta x}(\cdot, t) \leq V$ for all $x \geq r_0$, $t < T$. Then*

$$|\epsilon_1^j(\Delta x, \varphi, \mathbf{a})| \leq V \frac{diam\,(spt\varphi)}{L} \Delta x \|\varphi\|_{\infty}, \qquad (9.20)$$

and

$$|\epsilon_1(\Delta x, \varphi, \mathbf{a})| \leq \frac{V\Lambda}{L} (diam\,(spt\varphi))^2 \|\varphi\|_{\infty}. \qquad (9.21)$$

Proof: Since $[u_{\Delta x}](x, t_j)$ is bounded by the sum of the wave strengths from $x_{i-\frac{1}{2}}$ to $x_{i+\frac{1}{2}}$ for each x at time $t = t_j$, it follows that

$$|\epsilon_1^j| \leq \|\varphi\|_{\infty} \sum_{i,p} \|\gamma_{ij}^p\|_u \Delta x \leq \|\varphi\|_{\infty} V \frac{diam\,(spt\varphi)}{L} \Delta x, \qquad (9.22)$$

where $||\gamma||_u$ denotes the strength of a wave in u-space. This verifies (9.20). Consequently, if J is the smallest j so that $t = t_j$ upper bounds the support of φ, then $J = T/\Delta t$, where $T = J\Delta t$, and

$$|\epsilon_1| \leq \sum_{j=1}^{J} |\epsilon_1^j| \leq \frac{T}{\Delta t} \|\varphi\|_\infty \Delta x V \frac{diam\,(spt\varphi)}{L}$$

$$\leq \frac{V\Lambda}{L} (diam\,(spt\varphi))^2 \|\varphi\|_\infty$$

where $\Delta x / \Delta t \leq \Lambda$. □

We next show that ϵ_1^j, when taken as a function of a_j, has mean zero.

Lemma 6 *For approximate solutions* $u_{\Delta x}$,

$$\int_0^1 \epsilon_1^j \, da_j = 0 \tag{9.23}$$

Proof: The proof follows from Fubini's theorem.

$$\int_0^1 \epsilon_1^j \, da_j = \int_0^1 \sum_0^\infty \int_{x_{i-\frac{1}{2}}}^{x_{i+\frac{1}{2}}} [u_{\Delta x, \mathbf{a}}(x_i + a_j \Delta x, t_j) - u_{\Delta x, \mathbf{a}}(x, t_j)] \, dx \, da_j$$

$$= \sum_{i=0}^\infty \left\{ \int_{x_{i-\frac{1}{2}}}^{x_{i+\frac{1}{2}}} \int_0^1 u_{\Delta x, \mathbf{a}}(x_i + a_j \Delta x, t_j) \, da_j \, dx \right.$$

$$\left. - \int_0^1 \int_{x_{i-\frac{1}{2}}}^{x_{i+\frac{1}{2}}} u_{\Delta x, \mathbf{a}}(x, t_j) \, dx \, da_j \right\}$$

$$= 0,$$

which was to be proved. (Here we used $u_{\Delta x, \mathbf{a}}$ to express the dependence of the approximate solution $u_{\Delta x}$ on the sample sequence $\mathbf{a}$.) □

We now show that the functions ϵ_1^j are orthogonal, when taken as elements of $L^2(\Pi)$.

Lemma 7 *Suppose φ has compact support, and is piecewise constant on rectangles $\mathcal{R}_{ij}$. Then if $j_1 \neq j_2$, we have $\epsilon_1^{j_1} \perp \epsilon_1^{j_2}$ where orthogonality is with respect to the inner product on $L^2(\Pi)$.*

Proof: Using Lemma 7 in calculating the inner product

$$\langle \epsilon_1^{j_1}, \epsilon_1^{j_2} \rangle = \int \epsilon_1^{j_1} \epsilon_1^{j_2} (\Pi \, da_j) = \int \left(\int \epsilon_1^{j_1} \epsilon_1^{j_2} \, da_{j_2} \right) \Pi_{j \neq j_2} \, da_j$$

$$= \int \epsilon_1^{j_1} \left(\int \epsilon_1^{j_2} \, da_{j_2} \right) \Pi_{j \neq j_2} \, da_j$$

$$= 0,$$

verifying orthogonality. □

It follows immediately from Lemma 7 that

$$\|\epsilon_1\|_2^2 = \sum_j \|\epsilon_1^j\|_2^2, \tag{9.24}$$

which we use in our next theorem to finally show that there is a subsequence so that $\epsilon_1 \to 0$ as $\Delta x \to 0$ for almost any $\mathbf{a} \in \mathbf{\Pi}$.

Theorem 6 *Suppose that $TV_{[x,x+L]}u_{\Delta x}(\cdot, t) \leq V$ for all $r_0 \leq x < \infty$, $0 \leq t < T$. Then there is a null set $N \subset \mathbf{\Pi}$ and a sequence Δx_k such that for all $\mathbf{a} \in \mathbf{\Pi} - N$ and $\varphi \in C_0^1(t > 0)$, we have $\epsilon_1(\Delta x, \varphi, \mathbf{a}) \to 0$ as $k \to \infty$.*

Proof: Combining (9.24) and (9.20), and using the fact that $\int_{\mathbf{\Pi}} da = 1$, we have

$$\begin{aligned}
\|\epsilon_1(\Delta x, \varphi, \mathbf{a})\|_2^2 &= \sum_j \|\epsilon_1^j(\Delta x, \varphi, \mathbf{a})\|_2^2 \\
&\leq \sum_j \|\epsilon_1^j(\Delta x, \varphi, \mathbf{a})\|_\infty^2 \\
&\leq \sum_{j=0}^J V^2(\Delta t_k)^2 \|\varphi\|_\infty^2 \frac{(diam\,(spt\varphi))^2}{L^2} \\
&\leq V^2(\Delta t_k) \frac{(diam\,(spt\varphi))^3}{L^2} \|\varphi\|_\infty^2,
\end{aligned}$$

and hence, for piecewise constant φ with compact support, there is a sequence $\Delta x_k \to 0$ such that $\epsilon_1 \to 0$ in L^2. If φ is continuous with compact support, then by (9.21),

$$\|\epsilon_1\|_2 \leq \|\epsilon_1\|_\infty \leq C\|\varphi\|_\infty. \tag{9.25}$$

Let $\{\varphi_l\}$ be a sequence of piecewise constant functions with constant support whose closure relative to the infinity norm contains the space of test functions that are continuous with compact support. For each l, there is a null set $N_l \subset \mathbf{\Pi}$ and a sequence $\Delta x_{k_n(l)} \to 0$ such that $\epsilon_1 \to 0$ pointwise, for all $a \in \mathbf{\Pi} - N_l$. Set $N = \bigcup_l N_l$, and let $a \in \mathbf{\Pi} - N$. By a diagonalization process, we can find a subsequence, Δx_k, such that for each l, $\epsilon_1 \to 0$ as $k \to \infty$. If ψ is any test function, then if $u \in \mathbf{\Pi} - N$, we have

$$\begin{aligned}
|\epsilon_1(\Delta x, \psi, \mathbf{a})| &\leq |\epsilon_1(\Delta x, \psi - \varphi_l, \mathbf{a})| + |\epsilon_1(\Delta x, \varphi_l, \mathbf{a})| \\
&\leq Const.\|\psi - \varphi_l\|_\infty + |\epsilon_1(\Delta x, \varphi_l, \mathbf{a})|.
\end{aligned}$$

It is now clear that given $\epsilon > 0$, there exists $N \in \mathbf{N}$ so that if $i, l \geq N$, then $|\epsilon_1(\Delta x, \psi, \mathbf{a})| \leq \epsilon$. □

References

[1] Thanks to Piotr Bizon for showing this to us in a Private Communication

[2] R. M. Colombo and N. H. Risebro, *Continuous dependence in the large for some hyperbolic conservation laws*, Comm. Partial Diff. Equat. 23 (1998), 1693-1718.

[3] R. Courant and K. Friedrichs, *Supersonic Flow and Shock–Waves*, Wiley-Interscience, 1948. 1972.

[4] G. Crasta and B. Piccoli, *Viscosity solutions and uniqueness for systems of inhomogeneous balance laws*, Discr. Cont. Dyn. Syst. 4 (1997), 477-502.

[5] B.A. Dubrovin, A.T. Fomenko and S.P.Novikov, *Modern Geometry-Methods and Applications*, Springer Verlag, 1984.

[6] A. Einstein, *Der feldgleichungen der gravitation*, Preuss. Akad. Wiss., Berlin, Sitzber. 1915b, pp. 844-847. **314**(1970), pp. 529-548.

[7] Lawrence C. Evans, *Partial Differential Equations, Vol 3A*, Berkeley Mathematics Lecture Notes, 1994 Preuss. Akad. Wiss., Berlin, Sitzber. 1915b, pp. 844-847. **314**(1970), pp. 529-548.

[8] J. Glimm, *Solutions in the large for nonlinear hyperbolic systems of equations*, Comm. Pure Appl. Math., **18**(1965), pp. 697-715.

[9] J. Groah, *Solutions of the Relativistic Euler equations in non-flat spacetimes*, (Thesis, UC-Davis).

[10] J. Groah and B. Temple, *A shock-wave formulation of the Einstein equations*, Methods and Applications of Analysis, **7**, No. 4,(2000), pp. 793-812.

[11] J. Groah and B. Temple, *A locally inertial Glimm scheme for General Relativity*, Seventh Workshop on Partial Differential Equations, Matematica Contemporanea, **18**(2002), pp. 163-179, edited by P. Dias, D. Marchesin, A. Nachbin, Carlos Tomei.

[12] S.W. Hawking and G.F.R. Ellis, *The Large Scale Structure of Spacetime*, Cambridge University Press, 1973.

[13] W. Israel, *Singular hypersurfaces and thin shells in general relativity*, IL Nuovo Cimento, Vol. XLIV B, N. 1, 1966, 1-14.

REFERENCES

[14] P.D. Lax, *Hyperbolic systems of conservation laws, II*, Comm. Pure Appl. Math., **10**(1957), pp. 537–566.

[15] P.D. Lax, *Shock–waves and entropy*. In: Contributions to Nonlinear Functional Analysis, ed. by E. Zarantonello, Academic Press, 1971, pp. 603-634.

[16] M. Luskin, B. Temple, *The existence of a global weak solution to the non-linear waterhammer problem*, Comm. Pure Appl. Math., **35**(1982), pp. 697-735.

[17] T. Makino, K. Mizohata and S. Ukai, *Global weak solutions of the relativistic Euler equations with spherical symmetry*, Japan J. Ind. and Appl. Math., Vol. 14, No. 1, 125-157 (1997).

[18] D. Marchesin and P.J. Paes-Leme, *A Riemann problem in gas dynamics with bifurcation*, PUC Report MAT 02/84, 1984.

[19] C. Misner, K. Thorne, and J. Wheeler, *Gravitation*, Freeman, 1973.

[20] T. Nishida, *Global solution for an initial boundary value problem of a quasilinear hyperbolic system*, Proc. Jap. Acad., **44**(1968), pp. 642-646.

[21] T. Nishida and J. Smoller *Solutions in the large for some nonlinear hyperbolic conservation laws*, Comm. Pure Appl. Math., **26**(1973), pp. 183-200.

[22] J.R. Oppenheimer and J.R. Snyder, *On continued gravitational contraction*, Phys. Rev., **56** 1939, pp. 455-459.

[23] J. Smoller, *Shock-Waves and Reaction-Diffusion Equations*, Springer Verlag, 1983.

[24] J. Smoller and B. Temple *Global solutions of the relativistic Euler equations*, Comm. Math. Phys., **157**(1993), p. 67-99.

[25] J. Smoller and B. Temple, *Shock–wave solutions of the Einstein equations: the Oppenheimer-Snyder model of gravitational collapse extended to the case of non-zero pressure*, Arch. Rat. Mech. Anal., **128** (1994), pp 249-297.

[26] J. Smoller and B. Temple, *Astrophysical shock–wave solutions of the Einstein equations*, Phys. Rev. D, **51**, No. 6 (1995).

[27] J. Smoller and B. Temple, *Shock-wave solutions in closed form and the Oppenheimer-Snyder limit in general relativity*, with J. Smoller, SIAM J. Appl. Math, Vol.58, No. 1, pp. 15-33, February, 1998.

[28] J. Smoller and B. Temple, *Cosmology with a shock wave*, Commun. Math. Phys. 210, 275-308 (2000).

[29] R.M. Wald, *General Relativity*, University of Chicago Press, 1984.

[30] S. Weinberg, *Gravitation and Cosmology: Principles and Applications of the General Theory of Relativity*, John Wiley & Sons, New York, 1972.

Editorial Information

To be published in the *Memoirs*, a paper must be correct, new, nontrivial, and significant. Further, it must be well written and of interest to a substantial number of mathematicians. Piecemeal results, such as an inconclusive step toward an unproved major theorem or a minor variation on a known result, are in general not acceptable for publication. Papers appearing in *Memoirs* are generally longer than those appearing in *Transactions*, which shares the same editorial committee.

As of August 1, 2004, the backlog for this journal was approximately 5 volumes. This estimate is the result of dividing the number of manuscripts for this journal in the Providence office that have not yet gone to the printer on the above date by the average number of monographs per volume over the previous twelve months, reduced by the number of volumes published in four months (the time necessary for preparing a volume for the printer). (There are 6 volumes per year, each containing at least 4 numbers.)

A Consent to Publish and Copyright Agreement is required before a paper will be published in the *Memoirs*. After a paper is accepted for publication, the Providence office will send a Consent to Publish and Copyright Agreement to all authors of the paper. By submitting a paper to the *Memoirs*, authors certify that the results have not been submitted to nor are they under consideration for publication by another journal, conference proceedings, or similar publication.

Information for Authors

Memoirs are printed from camera copy fully prepared by the author. This means that the finished book will look exactly like the copy submitted.

The paper must contain a *descriptive title* and an *abstract* that summarizes the article in language suitable for workers in the general field (algebra, analysis, etc.). The *descriptive title* should be short, but informative; useless or vague phrases such as "some remarks about" or "concerning" should be avoided. The *abstract* should be at least one complete sentence, and at most 300 words. Included with the footnotes to the paper should be the 2000 *Mathematics Subject Classification* representing the primary and secondary subjects of the article. The classifications are accessible from www.ams.org/msc/. The list of classifications is also available in print starting with the 1999 annual index of *Mathematical Reviews*. The Mathematics Subject Classification footnote may be followed by a list of *key words and phrases* describing the subject matter of the article and taken from it. Journal abbreviations used in bibliographies are listed in the latest *Mathematical Reviews* annual index. The series abbreviations are also accessible from www.ams.org/publications/. To help in preparing and verifying references, the AMS offers MR Lookup, a Reference Tool for Linking, at www.ams.org/mrlookup/. When the manuscript is submitted, authors should supply the editor with electronic addresses if available. These will be printed after the postal address at the end of the article.

Electronically prepared manuscripts. The AMS encourages electronically prepared manuscripts, with a strong preference for $\mathcal{A}_{\mathcal{M}}\mathcal{S}$-LaTeX. To this end, the Society has prepared $\mathcal{A}_{\mathcal{M}}\mathcal{S}$-LaTeX author packages for each AMS publication. Author packages include instructions for preparing electronic manuscripts, the *AMS Author Handbook*, samples, and a style file that generates the particular design specifications of that publication series. Though $\mathcal{A}_{\mathcal{M}}\mathcal{S}$-LaTeX is the highly preferred format of TeX, author packages are also available in $\mathcal{A}_{\mathcal{M}}\mathcal{S}$-TeX.

Authors may retrieve an author package from e-MATH starting from
`www.ams.org/tex/` or via FTP to `ftp.ams.org` (login as `anonymous`, enter
username as password, and type `cd pub/author-info`). The *AMS Author Handbook* and the *Instruction Manual* are available in PDF format following the author
packages link from `www.ams.org/tex/`. The author package can be obtained free
of charge by sending email to `pub@ams.org` (Internet) or from the Publication
Division, American Mathematical Society, 201 Charles St., Providence, RI 02904,
USA. When requesting an author package, please specify $\mathcal{A}_{\mathcal{M}}\mathcal{S}$-LaTeX or $\mathcal{A}_{\mathcal{M}}\mathcal{S}$-TeX, Macintosh or IBM (3.5) format, and the publication in which your paper will
appear. Please be sure to include your complete mailing address.

Sending electronic files. After acceptance, the source file(s) should be sent to
the Providence office (this includes any TeX source file, any graphics files, and the
DVI or PostScript file).

Before sending the source file, be sure you have proofread your paper carefully.
The files you send must be the EXACT files used to generate the proof copy that was
accepted for publication. For all publications, authors are required to send a printed
copy of their paper, which exactly matches the copy approved for publication, along
with any graphics that will appear in the paper.

TeX files may be submitted by email, FTP, or on diskette. The DVI file(s) and
PostScript files should be submitted only by FTP or on diskette unless they are
encoded properly to submit through email. (DVI files are binary and PostScript
files tend to be very large.)

Electronically prepared manuscripts can be sent via email to
`pub-submit@ams.org` (Internet). The subject line of the message should include
the publication code to identify it as a Memoir. TeX source files, DVI files, and
PostScript files can be transferred over the Internet by FTP to the Internet node
`e-math.ams.org` (130.44.1.100).

Electronic graphics. Comprehensive instructions on preparing graphics are available at `www.ams.org/jourhtml/graphics.html`. A few of the major requirements are given here.

Submit files for graphics as EPS (Encapsulated PostScript) files. This includes
graphics originated via a graphics application as well as scanned photographs or
other computer-generated images. If this is not possible, TIFF files are acceptable
as long as they can be opened in Adobe Photoshop or Illustrator. No matter what
method was used to produce the graphic, it is necessary to provide a paper copy to
the AMS.

Authors using graphics packages for the creation of electronic art should also
avoid the use of any lines thinner than 0.5 points in width. Many graphics packages
allow the user to specify a "hairline" for a very thin line. Hairlines often look
acceptable when proofed on a typical laser printer. However, when produced on a
high-resolution laser imagesetter, hairlines become nearly invisible and will be lost
entirely in the final printing process.

Screens should be set to values between 15% and 85%. Screens which fall outside
of this range are too light or too dark to print correctly. Variations of screens within
a graphic should be no less than 10%.

Inquiries. Any inquiries concerning a paper that has been accepted for publication should be sent directly to the Electronic Prepress Department, American
Mathematical Society, 201 Charles St., Providence, RI 02904, USA.

Editors

This journal is designed particularly for long research papers, normally at least 80 pages in length, and groups of cognate papers in pure and applied mathematics. Papers intended for publication in the *Memoirs* should be addressed to one of the following editors. In principle the Memoirs welcomes electronic submissions, and some of the editors, those whose names appear below with an asterisk (*), have indicated that they prefer them. However, editors reserve the right to request hard copies after papers have been submitted electronically. Authors are advised to make preliminary email inquiries to editors about whether they are likely to be able to handle submissions in a particular electronic form.

*Algebra to ROBERT GURALNICK, Department of Mathematics, University of Southern California, Los Angeles, CA 90089-1113; email: guralnic@math.usc.edu

Algebraic geometry to DAN ABRAMOVICH, Department of Mathematics, Boston University, 111 Cummington St., Boston, MA 02215; email: abramovic@bu.edu

*Algebraic number theory to V. KUMAR MURTY, Department of Mathematics, University of Toronto, 100 St. George Street, Toronto, ON M5S 1A1, Canada; email: murty@math.toronto.edu

Algebraic topology and cohomology of groups to STEWART PRIDDY, Department of Mathematics, Northwestern University, 2033 Sheridan Road, Evanston, IL 60208-2730; email: priddy@math.nwu.edu

Combinatorics and Lie theory to SERGEY FOMIN, Department of Mathematics, University of Michigan, Ann Arbor, Michigan 48109-1109; email: fomin@umich.edu

Complex analysis and complex geometry to DUONG H. PHONG, Department of Mathematics, Columbia University, 2990 Broadway, New York, NY 10027-0029; email: phong@math.columbia.edu

*Differential geometry and global analysis to LISA C. JEFFREY, Department of Mathematics, University of Toronto, 100 St. George St., Toronto, ON Canada M5S 3G3; email: jeffrey@math.toronto.edu

Dynamical systems and ergodic theory to ROBERT F. WILLIAMS, Department of Mathematics, University of Texas, Austin, Texas 78712-1082; email: bob@math.utexas.edu

*Functional analysis and operator algebras to MARIUS DADARLAT, Department of Mathematics, Purdue University, 150 N. University St., West Lafayette, IN 47907-2067; email: mdd@math.purdue.edu

*Geometric analysis to TOBIAS COLDING, Courant Institute, New York University, 251 Mercer St., New York, NY 10012; email: colding@cims.nyu.edu

*Geometric analysis to MLADEN BESTVINA, Department of Mathematics, University of Utah, 155 South 1400 East, JWB 233, Salt Lake City, Utah 84112-0090; email: bestvina@math.utah.edu

Harmonic analysis to ALEXANDER NAGEL, Department of Mathematics, University of Wisconsin, 480 Lincoln Drive, Madison, WI 53706-1313; email: nagel@math.wisc.edu

Harmonic analysis, representation theory, and Lie theory to ROBERT J. STANTON, Department of Mathematics, The Ohio State University, 231 West 18th Avenue, Columbus, OH 43210-1174; email: stanton@math.ohio-state.edu

*Logic to STEFFEN LEMPP, Department of Mathematics, University of Wisconsin, 480 Lincoln Drive, Madison, Wisconsin 53706-1388; email: lempp@math.wisc.edu

Number theory to HAROLD G. DIAMOND, Department of Mathematics, University of Illinois, 1409 W. Green St., Urbana, IL 61801-2917; email: diamond@math.uiuc.edu

*Ordinary differential equations, and applied mathematics to PETER W. BATES, Department of Mathematics, Michigan State University, East Lansing, MI 48824-1027; email: peter@math.msu.edu

*Partial differential equations to PATRICIA E. BAUMAN, Department of Mathematics, Purdue University, West Lafayette, IN 47907-1395; email: bauman@math.purdue.edu

*Probability and statistics to KRZYSZTOF BURDZY, Department of Mathematics, University of Washington, Box 354350, Seattle, Washington 98195-4350; email: burdzy@math.washington.edu

*Real analysis and partial differential equations to DANIEL TATARU, Department of Mathematics, University of California, Berkeley, Berkeley, CA 94720; email: tataru@ math.berkeley.edu

All other communications to the editors should be addressed to the Managing Editor, WILLIAM BECKNER, Department of Mathematics, University of Texas, Austin, TX 78712-1082; email: beckner@math.utexas.edu.

Titles in This Series

815 **Martin Bendersky and Donald M. Davis,** V_1-periodic homotopy groups of $SO(n)$, 2004

814 **Johannes Huebschmann,** Kähler spaces, nilpotent orbits, and singular reduction, 2004

813 **Jeff Groah and Blake Temple,** Shock-wave solutions of the Einstein equations with perfect fluid sources: Existence and consistency by a locally inertial Glimm scheme, 2004

812 **Richard D. Canary and Darryl McCullough,** Homotopy equivalences of 3-manifolds and deformation theory of Kleinian groups, 2004

811 **Ottmar Loos and Erhard Neher,** Locally finite root systems, 2004

810 **W. N. Everitt and L. Markus,** Infinite dimensional complex symplectic spaces, 2004

809 **J. T. Cox, D. A. Dawson, and A. Greven,** Mutually catalytic super branching random walks: Large finite systems and renormalization analysis, 2004

808 **Hagen Meltzer,** Exceptional vector bundles, tilting sheaves and tilting complexes for weighted projective lines, 2004

807 **Carlos A. Cabrelli, Christopher Heil, and Ursula M. Molter,** Self-similarity and multiwavelets in higher dimensions, 2004

806 **Spiros A. Argyros and Andreas Tolias,** Methods in the theory of hereditarily indecomposable Banach spaces, 2004

805 **Philip L. Bowers and Kenneth Stephenson,** Uniformizing dessins and Belyĭ maps via circle packing, 2004

804 **A. Yu Ol'shanskii and M. V. Sapir,** The conjugacy problem and Higman embeddings, 2004

803 **Michael Field and Matthew Nicol,** Ergodic theory of equivariant diffeomorphisms: Markov partitions and stable ergodicity, 2004

802 **Martin W. Liebeck and Gary M. Seitz,** The maximal subgroups of positive dimension in exceptional algebraic groups, 2004

801 **Fabio Ancona and Andrea Marson,** Well-posedness for general 2×2 systems of conservation law, 2004

800 **V. Poénaru and C. Tanas,** Equivariant, almost-arborescent representation of open simply-connected 3-manifolds; A finiteness result, 2004

799 **Barry Mazur and Karl Rubin,** Kolyvagin systems, 2004

798 **Benoît Mselati,** Classification and probabilistic representation of the positive solutions of a semilinear elliptic equation, 2004

797 **Ola Bratteli, Palle E. T. Jorgensen, and Vasyl' Ostrovs'kyĭ,** Representation theory and numerical AF-invariants, 2004

796 **Marc A. Rieffel,** Gromov-Hausdorff distance for quantum metric spaces/Matrix algebras converge to the sphere for quantum Gromov-Hausdorff distance, 2004

795 **Adam Nyman,** Points on quantum projectivizations, 2004

794 **Kevin K. Ferland and L. Gaunce Lewis, Jr.,** The $RO(G)$-graded equivariant ordinary homology of G-cell complexes with even-dimensional cells for $G = \mathbb{Z}/p$, 2004

793 **Jindřich Zapletal,** Descriptive set theory and definable forcing, 2004

792 **Inmaculada Baldomá and Ernest Fontich,** Exponentially small splitting of invariant manifolds of parabolic points, 2004

791 **Eva A. Gallardo-Gutiérrez and Alfonso Montes-Rodríguez,** The role of the spectrum in the cyclic behavior of composition operators, 2004

790 **Thierry Lévy,** Yang-Mills measure on compact surfaces, 2003

789 **Helge Glöckner,** Positive definite functions on infinite-dimensional convex cones, 2003

788 **Robert Denk, Matthias Hieber, and Jan Prüss,** $\mathcal{R}$-boundedness, Fourier multipliers and problems of elliptic and parabolic type, 2003

TITLES IN THIS SERIES

787 **Michael Cwikel, Per G. Nilsson, and Gideon Schechtman,** Interpolation of weighted Banach lattices/A characterization of relatively decomposable Banach lattices, 2003

786 **Arnd Scheel,** Radially symmetric patterns of reaction-diffusion systems, 2003

785 **R. R. Bruner and J. P. C. Greenlees,** The connective K-theory of finite groups, 2003

784 **Desmond Sheiham,** Invariants of boundary link cobordism, 2003

783 **Ethan Akin, Mike Hurley, and Judy A. Kennedy,** Dynamics of topologically generic homeomorphisms, 2003

782 **Masaaki Furusawa and Joseph A. Shalika,** On central critical values of the degree four L-functions for GSp(4): The Fundamental Lemma, 2003

781 **Marcin Bownik,** Anisotropic Hardy spaces and wavelets, 2003

780 **S. Marmi and D. Sauzin,** Quasianalytic monogenic solutions of a cohomological equation, 2003

779 **Hansjörg Geiges,** h-principles and flexibility in geometry, 2003

778 **David B. Massey,** Numerical control over complex analytic singularities, 2003

777 **Robert Lauter,** Pseudodifferential analysis on conformally compact spaces, 2003

776 **U. Haagerup, H. P. Rosenthal, and F. A. Sukochev,** Banach embedding properties of non-commutative L^p-spaces, 2003

775 **P. Lochak, J.-P. Marco, and D. Sauzin,** On the splitting of invariant manifolds in multidimensional near-integrable Hamiltonian systems, 2003

774 **Kai A. Behrend,** Derived ℓ-adic categories for algebraic stacks, 2003

773 **Robert M. Guralnick, Peter Müller, and Jan Saxl,** The rational function analogue of a question of Schur and exceptionality of permutation representations, 2003

772 **Katrina Barron,** The moduli space of $N = 1$ superspheres with tubes and the sewing operation, 2003

771 **Shigenori Matsumoto,** Affine flows on 3-manifolds, 2003

770 **W. N. Everitt and L. Markus,** Elliptic partial differential operators and symplectic algebra, 2003

769 **Jie Wu,** Homotopy theory of the suspensions of the projective plane, 2003

768 **R. Höpfner and E. Löcherbach,** Limit theorems for null recurrent Markov processes, 2003

767 **Po Hu,** S-modules in the category of schemes, 2003

766 **Su Gao and Alexander S. Kechris,** On the classification of Polish metric spaces up to isometry, 2003

765 **Robert Bieri and Ross Geoghegan,** Connectivity properties of group actions on non-positively curved spaces, 2003

764 **J. Spandaw,** Noether-Lefschetz problems for degeneracy loci, 2003

763 **Yasuyuki Kachi and Eiichi Sato,** Segre's reflexivity and an inductive characterization os hyperquadrics, 2002

762 **Leiba Rodman, Ilya M. Spitkovsky, and Hugo Woerdeman,** Abstract band method via factorization, positive and band extensions of multivariable almost periodic matrix functions, and spectral estimation, 2002

761 **Oliver Druet and Emmanuel Hebey,** The AB program in geometric analysis : Sharp Sobolev inequalities and related problems, 2002

For a complete list of titles in this series, visit the
AMS Bookstore at **www.ams.org/bookstore/**.